ANNALS *of* THE NEW YORK ACADEMY OF SCIENCES

EDITOR-IN-CHIEF
Douglas Braaten

ASSOCIATE EDITOR
Rebecca E. Cooney

PROJECT MANAGER
Steven E. Bohall

EDITORIAL ADMINISTRATOR
Daniel J. Becker

Artwork and design by Ash Ayman Shairzay

The New York Academy of Sciences
7 World Trade Center
250 Greenwich Street, 40th Floor
New York, NY 10007-2157

annals@nyas.org
www.nyas.org/annals

THE NEW YORK ACADEMY OF SCIENCES BOARD OF GOVERNORS
SEPTEMBER 2011 - SEPTEMBER 2012

CHAIR
Nancy Zimpher

VICE-CHAIR
Kenneth L. Davis

TREASURER
Robert Catell

PRESIDENT
Ellis Rubinstein [ex officio]

SECRETARY
Larry Smith [ex officio]

CHAIRMAN EMERITUS
Torsten N. Wiesel

HONORARY LIFE GOVERNORS
Karen E. Burke
Herbert J. Kayden
John F. Niblack

GOVERNORS
Len Blavatnik
Mary Brabeck
Nancy Cantor
Martin Chalfie
Milton L. Cofield
Robin L. Davisson
Mikael Dolsten
Elaine Fuchs
Jay Furman
Alice P. Gast
Brian Greene
Thomas L. Harrison
Steven Hochberg
Toni Hoover
Thomas Campbell Jackson
John E. Kelly III
Mehmood Khan
Jeffrey D. Sachs
Kathe Sackler
Mortimer D.A. Sackler
John E. Sexton
George E. Thibault
Paul Walker
Anthony Welters
Frank Wilczek
Michael Zigman

INTERNATIONAL GOVERNORS
Seth F. Berkley
Manuel Camacho Solis
Gerald Chan
S. "Kris" Gopalakrishnan
Rajendra K. Pachauri
Russell Read
Paul Stoffels

EDITORIAL ADVISORY BOARD

**Ralph Adolphs**
California Institute of Technology

**Michael C. Corballis**
University of Auckland

**Michael Gazzaniga**
University of California, Santa Barbara

**Mel Goodale**
Western University

**Scott T. Grafton**
University of California, Santa Barbara

**Glyn Humphreys**
University of Birmingham

**Elisabetta Làdavas**
University of Bologna

**Adrian M. Owen**
University of Cambridge

**Elizabeth Phelps**
New York University

**Daniel Schacter**
Harvard University

Published by Blackwell Publishing
On behalf of the New York Academy of Sciences

Boston, Massachusetts
2012

ANNALS *of* THE NEW YORK ACADEMY OF SCIENCES

VOLUME 1251

ISSUE

# The Year in Cognitive Neuroscience

ISSUE EDITORS

Alan Kingstone[a] and Michael B. Miller[b]

[a]University of British Columbia and [b]University of California, Santa Barbara

TABLE OF CONTENTS



The New York Academy of Sciences believes it has a responsibility to provide an open forum for discussion of scientific questions. The positions taken by the authors and issue editors of *Annals of the New York Academy of Sciences* are their own and not necessarily those of the Academy unless specifically stated. The Academy has no intent to influence legislation by providing such forums.

## Become a Member Today of the New York Academy of Sciences

The New York Academy of Sciences is dedicated to identifying the next frontiers in science and catalyzing key breakthroughs. As has been the case for 200 years, many of the leading scientific minds of our time rely on the Academy for key meetings and publications that serve as the crucial forum for a global community dedicated to scientific innovation.

**Select one FREE *Annals* volume and up to five volumes for only $40 each.**

**Network and exchange ideas with the leaders of academia and industry.**

**Broaden your knowledge across many disciplines.**

**Gain access to exclusive online content.**

Join Online at **www.nyas.org**

Or by phone at **800.344.6902** (516.576.2270 if outside the U.S.).

Ann. N.Y. Acad. Sci. ISSN 0077-8923

ANNALS OF THE NEW YORK ACADEMY OF SCIENCES
Issue: *The Year in Cognitive Neuroscience*

# The role of strategies in motor learning

Jordan A. Taylor[1] and Richard B. Ivry[1,2]

[1]Department of Psychology, University of California, Berkeley, California. [2]Helen Wills Neuroscience Institute, University of California, Berkeley, California

Address for correspondence: Jordan A. Taylor, Ph.D., Department of Psychology, 3210 Tolman Hall, University of California, Berkeley, CA 94720-1650. jordan.a.taylor@berkeley.edu

**There has been renewed interest in the role of strategies in sensorimotor learning. The combination of new behavioral methods and computational methods has begun to unravel the interaction between processes related to strategic control and processes related to motor adaptation. These processes may operate on very different error signals. Strategy learning is sensitive to goal-based performance error. In contrast, adaptation is sensitive to prediction errors between the desired and actual consequences of a planned movement. The former guides what the desired movement should be, whereas the latter guides how to implement the desired movement. Whereas traditional approaches have favored serial models in which an initial strategy-based phase gives way to more automatized forms of control, it now seems that strategic and adaptive processes operate with considerable independence throughout learning, although the relative weight given the two processes will shift with changes in performance. As such, skill acquisition involves the synergistic engagement of strategic and adaptive processes.**

**Keywords:** motor learning; motor adaptation; motor skills; cognition

## Introduction

At a high school track meet in 1963, an athlete from Oregon changed the face of high jumping by falling, figuratively and literally, into a new technique.[1] Dick Fosbury had struggled to clear even modest heights using the "Western Roll," the popular technique at the time, in which the athlete extends his chest over the bar. After several embarrassing performances, Fosbury reverted to an antiquated scissors technique in which he simply hurdled sideways over the bar. On one attempt, he leaned back, thrusting his hips over the bar, and landing on his back. Not only did he clear the bar, but with subsequent jumps, he began to exaggerate this technique, fully throwing his back over the bar. By the end of the tournament, he had increased his personal record by half a foot. Although this improvement initially brought him up to the level achieved by top performers who were using the Western Roll, Fosbury went on to refine the technique over subsequent years, with his crowning achievement being a gold medal at the 1968 Olympics in Mexico City. Within a few years, nearly all jumpers had adopted the technique that to this day bears his name, the Fosbury Flop.

Fosbury's success led to a paradigm shift in the high jumping world. The impact of his technique was similar to that observed with other major innovations in high jumping (Fig. 1). The progression of world records generally shows a cyclical pattern. After the introduction of a new technique, there is a period in which the world record climbs in a steady manner over a relatively short period, followed by a rather lengthy plateau. Indeed, the current plateau of the Fosbury era has lasted since 1993, when Javier Sotomayor of Cuba cleared 2.45 m.

The history of high jumping establishes the theme for this review. When we think about motor skills, we typically focus on the performer's ability to execute a movement: how an exceptional quarterback has a rocket arm, or how the star tennis player gets such extraordinary power on her two-handed backhand. Missing from much of this discussion, however, is the role of insight and strategy. What led Fosbury to try going over the bar

doi: 10.1111/j.1749-6632.2011.06430.x

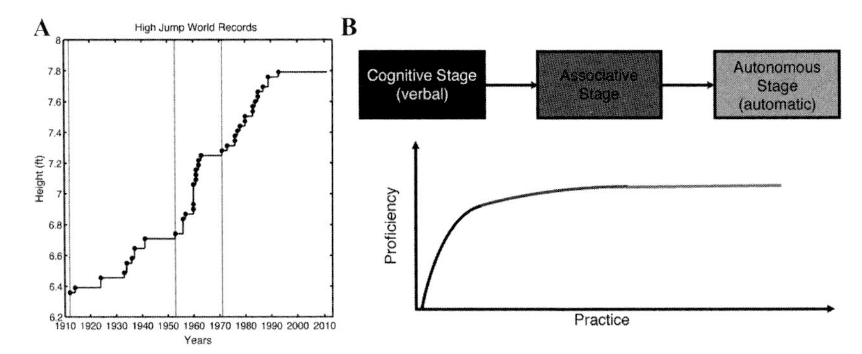


**Figure 1.** (A) World high jump records over the past century. Dashed vertical lines mark periods in which a particular technique was used by almost all jumpers. (B) Fitts and Posner model of skill acquisition. Their theory posits that skill acquisition follows three sequential stages: cognitive (black), associative (dark gray), and autonomous (light gray). The rate of skill acquisition varies across the three stages.

backward? How does the application of a cognitive strategy change performance and ultimately affect learning?

Studies of motor learning give little consideration to the role of cognitive strategies, in part because such processes are generally hard to formalize and are often variable. We address this limitation in this review and highlight experimental methods that have sought to directly assess the contribution of cognitive strategies in sensorimotor adaptation. We then discuss how computational models can incorporate such processes, and provide a means to understand quantitatively the contribution of cognitive processes to motor learning.

## Stages of learning

Fitts and Posner[2] proposed a model of skill acquisition that centered on three stages. In their now-classic theory, performance was characterized by three sequential stages, termed the cognitive, associative, and autonomous stages (Fig. 1B). The cognitive stage marks the period in which the task goals are established and used to determine the appropriate sequence of actions to achieve the desired goal. Learning at this stage generally involves the use of explicit knowledge. For Fosbury, the decision to go over the bar backward would constitute the cognitive stage. Once the action sequence has been determined, the learner enters the associative stage in which attention may be focused on specific details of the sequence, determining the appropriate subparts and transitions. This stage may require some exploration of the solution space, perhaps with one segment being overhauled to ensure that the overall action is executed in a smooth and coordinated manner. Although Fosbury pioneered the idea of leading with his back, other jumpers came along to refine this general strategy and develop the proper foot placement, timing, and body orientation. The final stage of learning is the autonomous stage, the phase in which the action is practiced to hone performance into an automatized routine. For high jumping, we might say that Fosbury and his peers guided a generation of jumpers through cognitive and associative stages. But each of these individuals must put in the countless hours of practice required for elite performance that results from the autonomous stage.

More generally, learning curves across a wide range of tasks show a general shape that conforms to the basic model of Fitts and Posner.[2] There is an initial phase marked by rapid improvements in performance, followed by a more gradual phase in which performance gains accrue much more slowly. Numerous theories have been proposed to account for these functions.[3,4] In the Fitts and Posner[2] model, the emphasis is on a shift in control in which initial, explicit control gives way to more routinized forms of control. Other models have emphasized that these functions may reflect the parallel operation of multiple processes. Logan[5] introduced a theory in which execution reflected a horse race between an algorithmic, explicit process (akin to the cognitive stage) and a memory-retrieval process. Although both processes were assumed to operate at all stages of performance, a shift in their relative contribution naturally arises over time as the memory base builds up.

Psychological theories such as those of Fitts and Posner[2] or Logan[5] offer a general framework for understanding skill acquisition functions. Similar learning functions are observed in studies of sensorimotor adaptation. This work has spawned a rich computational literature in which performance changes are analyzed from an engineering perspective grounded in ideas related to control systems. However, this new modeling perspective has just begun to address the role of cognitive processes during motor learning, processes that were inherent in the models of Fitts and Posner[2] and Logan.[5]

## Sensorimotor adaptation

A common method to study motor learning is to introduce a perturbation into the experimental context. Participants must learn to compensate for these perturbations to re-achieve a high level of performance. The perturbation introduces an error between a motor command and a desired outcome. This error signal serves as input used to update an internal model, a mapping between a desired goal and the motor response necessary to achieve that goal (Fig. 2A). In this manner, the mapping is refined to adjust the motor commands. In general, the goal is assumed to remain constant; for example, the high jumper always wants to clear the bar. Failure to achieve this goal may lead to changes in performance, such as a modification in the takeoff angle or timing of the initial thrust.

A wide range of experimental paradigms has been employed to study sensorimotor adaptation. One popular task involves a visuomotor rotation in which the visual feedback indicating hand position is perturbed (Fig. 2B). Visuomotor adaptations are common in everyday life. For example, using a computer mouse requires learning the mapping between the hand-held device and the position of a cursor on a computer screen. In the experimental context, this mapping can be perturbed. In many studies the input–output relationship between a device such as a mouse or joystick is altered. In other conditions, participants make reaching movements in which the hand is not visible, and a cursor is used to provide feedback. The natural mapping between the hand and space is distorted. In a visuomotor rotation, feedback of the hand position is adjusted in a rotational manner. The rotations typically take on values ranging between 30° and 60°.[6–9] This task is nicely situated to examine the interaction of action selection and motor execution.

Participants readily adapt to visual perturbations, showing a reduction in target errors with training. Adaptation proceeds in a gradual manner, in which the learning function typically conforms to an exponentially decaying function. This pattern is consistent with the hypothesis that an error signal is used to continuously adjust the visuomotor mapping, with the magnitude of the change proportional to the error. Thus, large errors observed early in

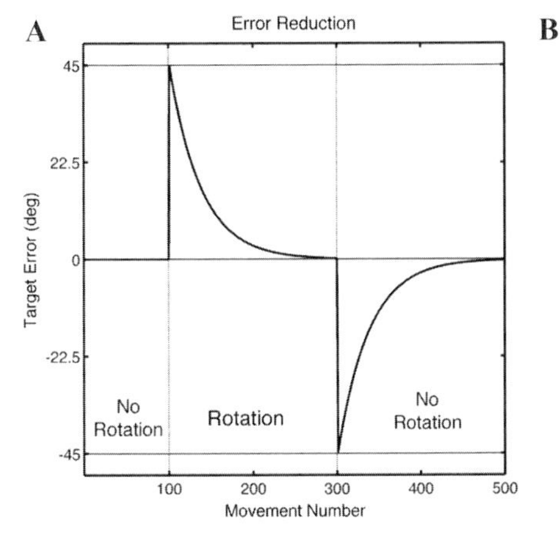


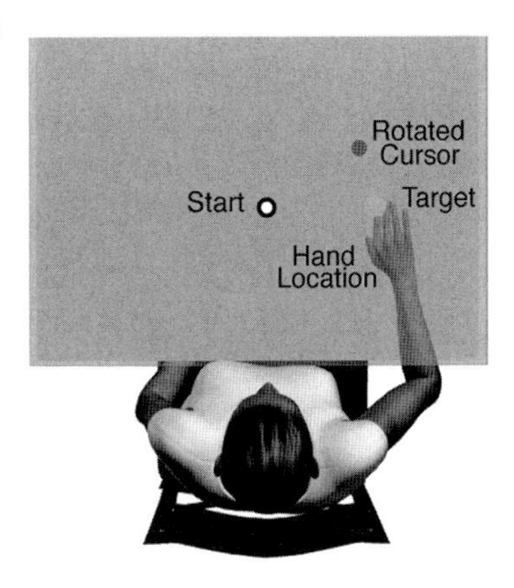


**Figure 2.** (A) Hypothetical learning curve during adaptation to visuomotor rotation. A 45° rotation is imposed during movements 81–160. Target errors are initially in the direction of the rotation, but with training, adaptation occurs. The rotation is removed on trial 161 and an aftereffect is observed in which target errors are in the direction opposite to the rotation. (B) Virtual reality environments are used to impose systematic transformation between actual and projected hand position. Vision of the limb is occluded. In this example, the target is the gray circle and a 45° rotation clockwise led to displacement of feedback location (black circle).

training produce relatively large changes in performance compared to the effects of small changes that occur late in training. After training, the visuomotor rotation is removed and the original environment is reinstated. This induces a pronounced aftereffect with errors now occurring in the opposite direction of the initial distortion. If feedback is provided, the learning process is repeated to "wash out" the effects of the altered sensorimotor mapping and restore the original mapping. The presence of an aftereffect is considered the hallmark of true adaptation. Performance gains (e.g., reduced error) are also possible from the implementation of a strategy or a change in the selected action; however, in either case, an aftereffect should either be absent or diminish rapidly. The term "motor learning" generally encompasses changes that may entail a combination of the alteration of a sensorimotor map from adaptation and performance gains resulting from other, nonadaptive processes.

## Movement strategies

In the typical visuomotor adaptation study, the perturbation is suddenly introduced after a baseline period of training. On the first trial, the participant will be surprised to see a large error. For example, if feedback is only presented at the endpoint of the movement, the participant suddenly sees feedback indicating an error of 30°. Although one trial may be written off as a chance event, the repetition of this error with subsequent trials leads many participants to become aware that the environment has been perturbed. This awareness suggests an alternative account of visuomotor adaptation: the participant may adopt a strategy to aim their movement in the direction opposite the rotation.

It is generally assumed that strategy-based learning is not a major contributing factor in visuomotor adaptation. First, the learning function during the washout period is similar in form, albeit with a steeper learning rate, than that observed during the initial learning phase. If the participant were employing a strategy, one would expect washout to occur in a more or less categorical manner. That is, the participant could simply choose to not apply the strategy to again establish the normal sensorimotor mapping. Nonetheless, although learning may involve more than the instantiation of a strategy, it is important to recognize that there may be a contribution to performance from strategic processes. More importantly, it is important to consider how a strategy, if employed, influences processes involved in sensorimotor adaptation.

As an everyday example, consider how people adjust their behavior when riding on a crowded city bus. If forced to stand, one might cocontract the leg muscles to increase stiffness, making the legs (and person) resistant to small and unexpected changes in acceleration. Although this strategy can be effective, it is energetically wasteful. An alternative, more adaptive procedure is for the motor system to predict upcoming perturbations. Suppose both processes are operative. How does the utilization of the cocontraction strategy influence motor adaptation? If we assume the input to the adaptation system is a motor error, the adaptation system may work in a suboptimal manner because the motor error is significantly reduced by cocontraction. Is the adaptation system able to incorporate information about the level of cocontraction in its computations? Or does this system operate in a modular manner, ignorant of the context created by the decision of the person to stiffen the limbs?

One approach to exploring these questions is to compare conditions in which participants are either aware or unaware of an experimentally induced perturbation.[10,11] As noted above, in the standard visuomotor adaptation task, a large rotation is abruptly imposed. This produces both a large error signal and, in many situations, creates a situation in which the participants are aware that the environment has been altered. Alternatively, the perturbation can be introduced in small increments, such as 1° every 10 trials, with the full 90° rotation only achieved after 300 trials (Fig. 3A). Under these conditions, participants generally have no awareness of the perturbation because the induced visual error is within the bounds of the variability associated with the motor system. Adaptation occurs in a continuous manner under these conditions, preventing the small errors from accumulating to a level that is noticeable (Fig. 3B). By the end of training, the performance of the participants is similar. However, when the rotation is switched off, the aftereffect is generally smaller for participants in the abrupt condition.[10] Moreover, when participants gain knowledge of the rotation through self-inference or instruction, performance is associated with large trial-by-trial variance and longer reaction times, at least in the early stages of adaptation.[12–14]

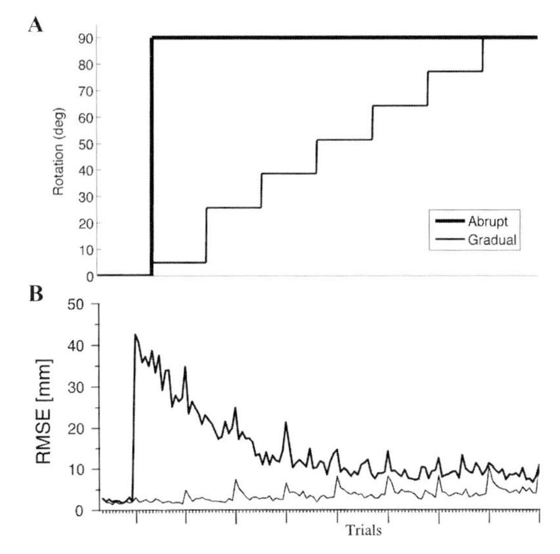


**Figure 3.** (A) A 90° visuomotor rotation can be introduced in a single step (thick line) or gradually across multiple steps (thin line). (B) Curvature of trajectories as measured by root mean square error (RMSE) when online feedback is available for the abrupt (thick) and gradual (thin) conditions. By the end of training, the degree of error is similar for the two conditions. Adapted from Kagerer *et al.*[10]

These features suggest that the participants' awareness of the perturbation in the abrupt condition may alter their performance. Some participants may opt to test strategies to offset the large error observed after the onset of the rotation,[15] although the use of such strategies is likely to be highly idiosyncratic across individuals, as well as variable across trials for a given individual. Consistent with a strategy-use hypothesis is the observation that participants with higher spatial working memory capacity tend to exhibit faster rates of adaptation.[16] Verbal working memory has also been linked to motor adaptation. When sequentially trained with two different sensorimotor mappings, a subsequent verbal learning task only disrupted the memory of the most recent mapping.[17] Interestingly, this effect was not observed when the verbal task did not involve learning (e.g., vowel counting), suggesting that the point of overlap was within processes associated with learning *per se*. The specificity here may reflect interference with a verbal strategy (e.g., "On the next trial, I will push to the left").

Dual-task manipulations have offered an indirect method for assessing strategic contributions to motor learning. The underlying logic is that the cognitive requirements for maintaining and implementing a strategy would be taxed by a concurrent secondary task. Indeed, when a secondary task is performed concurrently with a sensorimotor adaptation task, performance gains during training are reduced.[18–21] Interestingly, even seemingly automatic processes, such as learning a new spatiotemporal walking pattern, are affected by dual-task interference.[22] Moreover, the effects of dual-task interference are not limited to conditions in which the participants are aware of the sensorimotor perturbation.[21] Thus, dual-task costs may be unrelated to the deployment of strategic processes. Instead, the interference may result from other shared stages of processing such as the sensory processing requirements for the primary and secondary tasks.

## Direct interaction of action selection and motor adaptation

In most experimental paradigms, the use of a strategy during motor learning is the prerogative of the participant; the experimenter can only infer strategy use from the behavior. Mazzoni and Krakauer[23] introduced a novel method to directly address the effect of strategy use on visuomotor adaptation. The workspace consisted of a display of eight visual landmarks, spaced 45° apart. On each trial, a visual target appeared at one of the landmarks. Participants were initially trained to reach directly to the target. After this baseline phase, a 45° counterclockwise rotation was introduced (Fig. 4A) and large visual errors were experienced for two trials. The experimenter then instructed the participant to use a strategy to counteract the rotation, aiming 45° in the clockwise direction to the neighboring landmark (Fig. 4B). The strategy was immediately effective, counteracting the visual error. Surprisingly, as training continued, performance deteriorated: the movement endpoints drifted over trials in the direction of the strategy. That is, the heading angles were greater than the instructed 45° (Fig. 4C). Thus, the participants' performance became worse with increasing practice (Fig. 4D).

What can account for this puzzling effect? Why would the system continue to change despite good on-target performance? Mazzoni and Krakauer[23] proposed that this phenomenon reflected the ongoing operation of an implicit motor adaptation system. Importantly, the error signal used for adaptation is based on the difference between the desired aiming location and the visual feedback of the

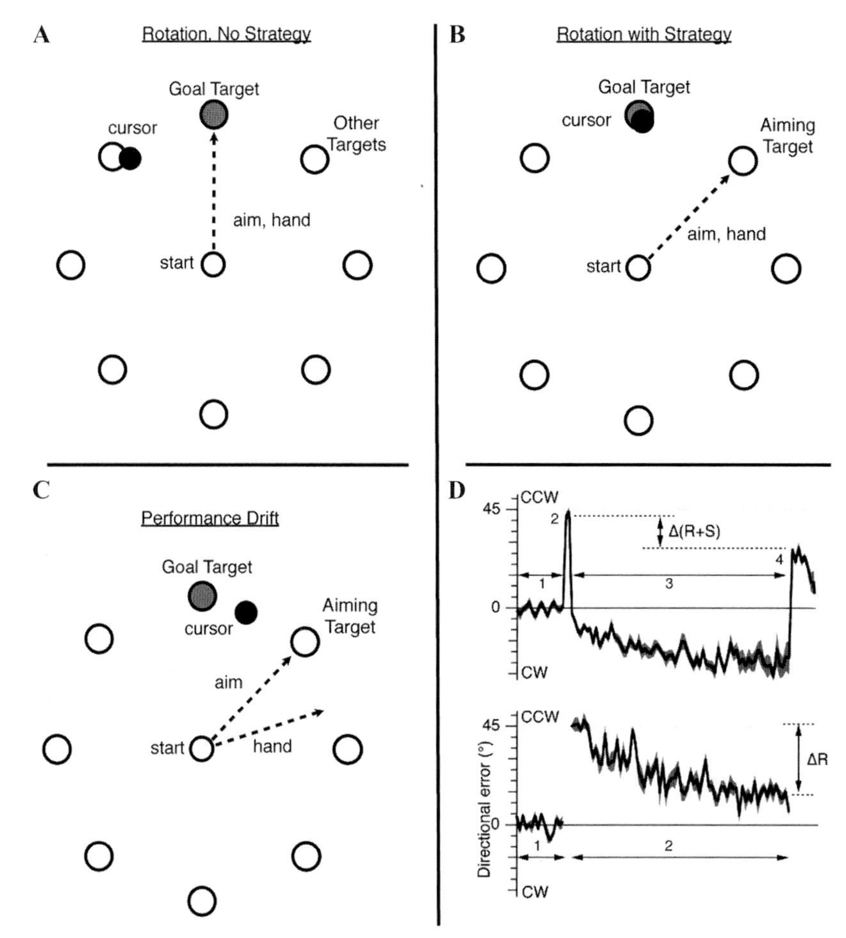


**Figure 4.** Visuomotor adaptation with landmarks. (A) Displays contain eight empty circles, indicating possible target locations arranged along an invisible ring. One circle turns gray, indicating the target for that trial. Feedback indicates the position of the cursor at the time the movement amplitude exceeds the radius of the ring. During rotation phases of the study, the feedback is presented 45° in the counterclockwise direction of the true hand position. (B) In the rotation with strategy block, participants were instructed to move to the neighboring target landmark to offset for the rotation. (C) By the end of rotation + strategy block, the reaches had drifted in the direction of the aiming target, resulting in increased target error. (D) Target errors are centered around zero during the baseline block (1). Large errors are observed when the rotation is unexpectedly introduced (2). When instructed to use the strategy, movements are initially very accurate but, over time, performance deteriorates with error drifting in the direction of the strategy (3). Aftereffect is observed when participants are instructed to stop using the strategy (4). Adapted from Mazzoni and Krakauer.[23]

cursor. That is, the adaptation process does not take into account the difference between the movement goal (e.g., the target) and movement feedback, despite the fact that this information defines task success. As emphasized by Mazzoni and Krakauer,[23] the drift phenomenon provides strong evidence that sensorimotor adaptation processes are segregated from goal-based movement strategies.

We set out to further explore the interaction, or lack thereof, between strategic and adaptation processes. In their original study, Mazzoni and Krakauer[23] limited training to 80 trials, and at the end of this period, the endpoint error had increased to over 25°, presumably because the adaptation system continued to compensate for the mismatch between the aiming location and the feedback location. Would this process continue to operate until the error was eliminated, resulting in an observed endpoint error of 45°? To test this, we quadrupled the training period to 320 trials.[24] With this extended training, the basic drift effect was observed over the first part of the training period, followed by

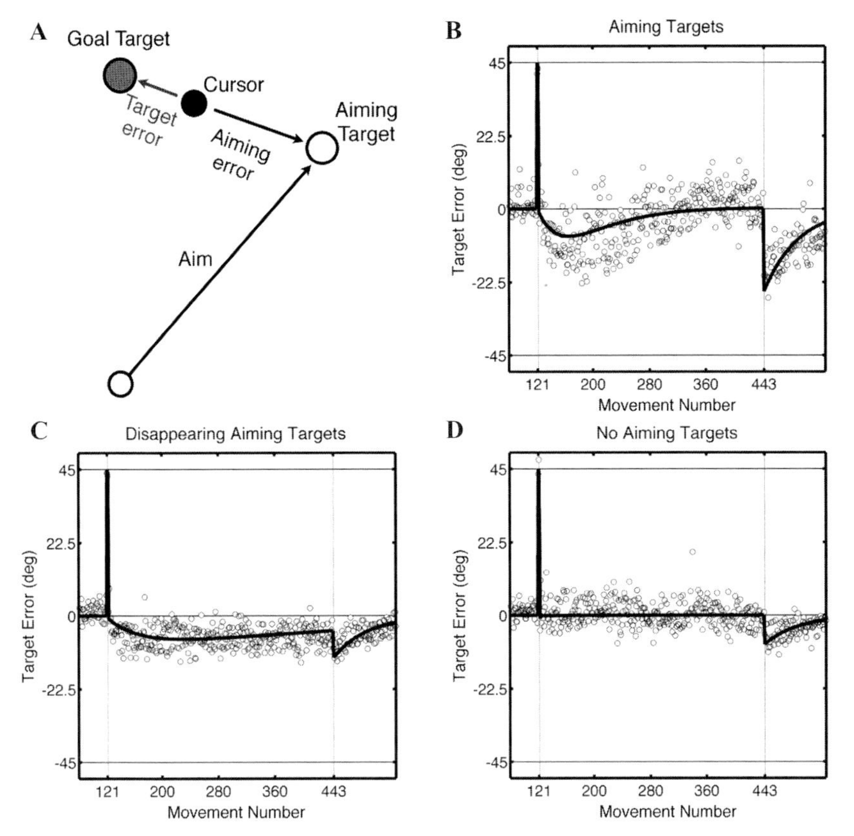


**Figure 5.** (A) Implicit adaptation is based on an aiming error signal (black), defined as the difference between the aiming location and feedback location. Strategy adjustment is based on target error (gray), the difference between the target location and feedback location. (B) Aiming targets present. When a strategy is implemented to offset a rotation, target error is initially small because performance is accurate. However, aiming error is large, and adaptation leads to deterioration in accuracy. When target error becomes large, the effect of strategy adjustment becomes more prominent, leading to a reversal of the drift. The system eventually stabilizes even though the two learning processes continue to operate throughout training. The aftereffect observed when the rotation is removed reveals the magnitude of implicit adaptation. Circles: observed data for the group. Solid curve: model fit. (C) Disappearing aiming target group. The aiming target was turned off at movement onset. (D) No aiming target group. The aiming targets were never visible.

a reversal, with performance becoming near perfect by the end of training.

To quantify this nonmonotonic behavior, we developed a novel state space model to capture trial-by-trial changes in performance.[24] The key idea in the model is that the output is the result of two learning processes, one associated with movement execution and the other with action planning. Moreover, these two processes use distinct error signals. First, the difference between the desired movement and the actual outcome, what we call aiming error, is used by the adaptation system to recalibrate an internal model. This component is similar to standard state space models of sensorimotor adaptation.[25–27] However, in most studies, the desired movement is usually directed at the target; in the strategy-use variant, the desired movement is now directed at an aiming location. Second, the difference between the target and feedback location defines a target error, a signal that is used to adjust the movement strategy (Fig. 5A).

By simulating the parallel operation of these two processes, the model produces an excellent match to the function produced by our participants. Implementing a strategy immediately offsets the rotation. Over time, the target error increases, drifting in the direction of the strategy, and then the function reverses to stabilize on correct performance (Fig. 5B). Importantly, the model does not entail any sort of "stages" in which control shifts between "strategy-based" and "adaptation-based" phases. Rather, both processes are always functional. Drift is prominent during the early phase of the training period after strategy implementation because the aiming error is large and target error is small. This results in large changes in the adaptation system and small adjustments of the strategy. As target error becomes large, strategy adjustments become more prominent. Importantly, even when performance stabilizes with minimal target error, the two processes continue to operate, achieving a stable tension. This tension is evident in the aftereffect observed when the rotation is removed and participants are told to stop using the strategy.

In addition to highlighting the parallel operation of two learning processes, the model also provides a fresh take on how strategic and adaptation processes interact. Consistent with the conjecture of Mazzoni and Krakauer,[23] the implicit adaptation system appears to be completely isolated from the strategy. The aiming error signal that is used to recalibrate this system to ensure accurate movement execution is based on the difference between the predicted and actual movement outcome; this signal does not incorporate the participant's strategy, leading to the paradoxical drift phenomenon. In contrast, the strategy is adjusted by an error signal that reflects the movement goal, which is defined as the difference between where the target is located and where the movement ended. Learning within this system modifies a representation relevant for action planning, focused on ensuring that the outcome of the action achieves the desired goal.

Although the adaptation system is modular in that it does not take into account the strategy (e.g., recognize that the feedback may be displaced because of a strategy adopted to negate a rotation), the impact of the adaptation system on strategy adjustment is unclear. In our model, the strategic process only has access to target error, the difference between the goal and the feedback; it does not have access to changes arising from adaptation. Thus, the influence is indirect.

As reviewed above, certain features of performance—slower RTs, increased variability, and smaller aftereffects—have led to the inference that participants may adopt a strategy to offset a large perturbation in visuomotor adaptation tasks. However, none of these studies has reported the drift phenomenon. This is puzzling if one assumes that by adopting a strategy (e.g., "aim in clockwise direction of target), a mismatch is created between the aiming location and the feedback location. Why should the explicit instruction in the use of a strategy produce drift, whereas self-discovery by the participant does not?

The answer seems to be because of a second important methodological difference between these standard visuomotor adaptation tasks and the variant introduced by Mazzoni and Krakauer.[23] The latter included visible landmarks spaced every 45° to provide a reference point for the strategy. In the standard task, these landmarks are absent; participants only see a stimulus at the target location. We propose that the landmarks serve as a proxy for the predicted location of the movement. That is, even though the adaptation system uses an error based on the difference between the predicted and actual movement, the landmarks provide a salient referent for the predicted movement location. When these landmarks are absent, the participant's sense of the predicted movement (e.g., 45° clockwise from the target) is likely uncertain, and thus the weighting given to the aiming error term is attenuated. We tested this idea by comparing conditions in which the landmarks were always present, disappeared at movement initiation, or were never presented (Figs. 5C and D). Consistent with the certainty hypothesis, the degree of drift was attenuated as uncertainty increased.[24] Indeed, when the landmarks were never present, drift was minimal throughout the training block.

## Reward-based learning and error-based adaptation

Although our modeling work entails two error terms, error-based learning may not be the most appropriate way to characterize the strategy adjustment process. Rather, learning within the strategic process, with its emphasis on the movement goal, may be better described in terms of models of reinforcement learning. These models are designed to account for how organisms explore different regions of a strategy space, attempting to identify the action policy that results in the largest reward. In our task, a shift in policy might occur when the rise in target error because of adaptation becomes too large. That is, when a chosen action fails to achieve the predicted reward, a new strategy might be adopted. In our study, only a few of the participants exhibited categorical-like changes in performance. For these participants, we observed abrupt reductions in target error, suggesting that the participant either stopped using the strategy or made a categorical change in their strategy (e.g., switched from aiming to a landmark to a position between two landmarks). For the other participants, the data suggest a more gradual change in the strategy with incremental changes that eventually led to movement success.

The relationship between reinforcement learning and sensorimotor adaptation was highlighted in a recent study in which participants were only provided with categorical feedback when learning a rotation.[28] In this format, the participants were not given visual feedback of their movement endpoint, but instead observed an explosion of the target on successful trials, those in which the hand passed within a criterion window. Compared to standard adaptation tasks, learning here was characterized by a dramatic increase in trial-by-trial variance, limited generalization to untrained movements, and reduced sensorimotor remapping. Furthermore, the learning function could be accounted for by a reward-based learning model in which action policies (aiming direction) were adjusted to maximize the rate of reward. Similar to the instructed strategy tasks, the Izawa and Shadmehr[28] study also suggests that motor learning may be composed of (at least) two processes: a reward-based action-selection process and an error-based adaptation process.

The form and certainty of the visual feedback influences the form of the representational changes associated with the performance changes. Aftereffects are larger when the feedback is continuous during adaptation, compared to conditions in which only endpoint feedback or knowledge of performance is provided.[29–31] Continuous visual feedback provides more salient spatiotemporal feedback of the relationship between the movement and feedback. As such, the information is more reliable, with a tight covariance between motor commands and their outcome. Increasing this covariance seems to facilitate adaptation as evident by the larger aftereffect. Conversely, decreasing this covariance promotes learning via strategic processes, evident in the attenuated aftereffect. Indeed, in their computational model, Izawa and Shadmehr[28] show how the uncertainty of visual feedback can distribute learning across adaptation or action-selection processes. As in our model, both processes seem to operate in parallel.

Movement errors have been thought to provide a signal that is used to incrementally update an internal model, one that predicts the consequences of motor commands given the dynamics of the environment. Generalization, in which movements to novel directions or in novel contexts show effects of learning, have been taken as evidence of the presence of an internal model for motor control. However, motor learning can arise by reinforcement, simply relating the success or failure of an executed movement. This form of learning has been called "model-free" because the feedback does not guide formation of an internal model of task dynamics, but rather only the value of potential actions or movements.[32] Model-free learning can also arise from pure repetition,[32,33] or what has been called use-dependent plasticity.[34] The work on model-based and model-free learning again emphasizes that motor learning is a composite term.

## Neural systems subserving action selection and motor execution

This two-process interpretation may also provide a new perspective for understanding the consequences of neural pathology on specific learning processes. Consider the effects of cerebellar damage on sensorimotor control and adaptation. Numerous studies have shown that patients with cerebellar ataxia show attenuated adaptation. These findings have been assumed to reflect the operation of a compromised learning system.[35–37] Alternatively, the patients' performance may reflect the implementation of a compensatory process, one in which they have come to rely on alternative forms of control. Lang and Bastian[38] observed that patients with cerebellar damage performed surprisingly well when asked to make rapid complex drawing movements, reaching a performance level comparable to that of control participants. However, when the drawing task was performed concurrently with a secondary task, their performance was markedly reduced. These results suggest that the patients may have relied on a strategy-based control system, one that was taxed by the inclusion of the secondary task.

We directly addressed this question by providing an explicit strategy to a group of individuals with bilateral cerebellar degeneration who were presented with a visuomotor rotation.[39] The patients had no difficulty implementing the strategy and were successful in immediately counteracting a 45° rotation. However, in comparison to control participants, the patients exhibited attenuated drift; their performance remained accurate across the training block. These results indicate that patients with cerebellar degeneration can use a strategy, and in fact, their use of a strategy remains stable because it is not disrupted by the operation of the adaptation system.

Our results are consistent with prior work showing that patients with cerebellar damage are impaired in sensorimotor adaptation.[35,36,40] It is interesting to ask why these individuals do not generate compensatory strategies in tasks using standard adaptation tasks. That is, if the error signal remains stable over trials (because of an impaired learning mechanism), why don't the patients come up with a strategy, given that they have little difficulty using one when given explicit instructions? It is possible that although the adaptation process is isolated from strategies, the reverse may not be true; adaptation processes may inform strategic processes. For example, visuomotor rotations create a complex pattern of errors. The adaptation system may aid in the formation of a movement strategy by providing detailed error information. Without this information, it may be difficult to generate a successful strategy because the strategy is designed to overcome these errors.

By this view, damage to the cerebellum may not only disrupt sensorimotor adaptation, but may also disrupt the generation of cognitive strategies. This hypothesis may also help to explain why patients with cerebellar degeneration show a greater impairment in learning to compensate for an abrupt, and large, force field perturbation, compared to when the perturbation is introduced gradually.[37] When errors are large, the cerebellum may work in concert with frontal areas to aid in an action-selection process required for the generation of a movement strategy.[41,42] Extensive reciprocal connections between the cerebellum and prefrontal cortex (PFC) are suggestive of a coordinated network that can integrate motor and cognitive processes.

Lesions of PFC, either virtually with transcranial magnetic stimulation or from naturally occurring lesions, can disrupt performance on sensorimotor learning tasks.[43] Patients with PFC lesions exhibit pronounced deficits in visuomotor adaptation.[44–46] Interestingly, these patients have difficulty describing the perturbation, or when aware of it, have difficulty reporting what action would be required to compensate for the perturbation.[44,45] In a similar vein, older adults show a slower rate of adaptation compared to younger individuals,[47] a deficit that is attenuated in older adults who are able to explicitly describe the perturbation.[48] Thus, this deficit may, in part, be related to a problem in generating strategies for motor adaptation.

## Generating a movement strategy

At present, considerable progress has been made in the development of computational models that describe sensorimotor adaptation. In contrast, we know little about the process of strategy development. What sources of information are used to generate and modify strategies? What are the dynamics of strategy change? Are there signals inherent in action or task-dependent variables that help define the solution space? What drove Fosbury to lean back when he reverted to the scissors technique? His willingness to break with the conventions of his day may have been fueled by the fact that the landing pit was now filled with shock absorbent foam rather than sawdust and wood chips. With softer materials, jumpers no longer had to worry about landing on their feet or hands.

To fully understand strategy development and change, it will be necessary to characterize the inputs to the strategy process. Popular tasks for studying action selection, such as the n-bandit task, are generally limited to a small, fixed set of discrete actions. This limits the search space to a finite set of action–outcome alternatives. In motor control, the search space is continuous and in some sense nearly infinite. It is also unclear if reward is the driving input for the strategy system. Reward in many contexts is discrete—a choice was either correct or incorrect. However, actions, especially when they involve complex sequences of movements, are much more varied. How would reward signals be used to train a strategic process with a nearly infinite action space? A reasonable experimental approach will require a more constrained situation, tasks in which there is a relatively limited set of potential actions to isolate the inputs and characterize the time course of the strategic process.

Work along these lines could provide a new perspective for understanding not only strategy change, but spontaneous strategy development.[49]

Considerable research has been devoted to understanding the processes underlying spontaneous insight in problem-solving tasks.[50,51] Borrowing techniques from the insight literature may offer clues to strategy development in the motor domain and help us understand progress in human performance. As shown in Figure 1, the world record in high jumping has not budged since 1993. Although we may see a new leaper perfect the Fosbury Flop and produce an incremental change in the record, the next ascent of the bar is likely to require the discovery of a radically new technique.

## Conflicts of interest

The authors declare no conflicts of interest.

## References

1. Hoffer, R. 2009. *Something in the Air: American Passion and Defiance in the 1968 Mexico City Olympics*. Free Press. New York.
2. Fitts, P.M. & M.I. Posner. 1967. *Human Performance*. Brooks/Cole Pub. Co. Belmont, CA.
3. Crossman, E.R.F.W. 1959. A theory of the acquisition of speed-skill. *Ergonomics* **2:** 143–166.
4. Seibel, R. 1963. Discrimination reaction time for a 1,023 alternative task. *J. Exp. Psychol.* **66:** 215–226.
5. Logan, G. 1988. Toward an instance theory of automatization. *Psychol. Rev.* **95:** 492–527.
6. Cunningham, H.A. 1989. Aiming error under transformed spatial mappings suggests a structure for visual-motor maps. *J. Exp. Psychol. Hum. Percept. Perform.* **15:** 493–506.
7. Imamizu, H., Y. Uno & M. Kawato. 1995. Internal representations of the motor apparatus: implications from generalization in visuomotor learning. *J. Exp. Psychol. Hum. Percept. Perform.* **21:** 1174–1198.
8. Pine, Z.M., J.W. Krakauer, J. Gordon & C. Ghez. 1996. Learning of scaling factors and reference axes for reaching movements. *Neuroreport* **7:** 2357–2361.
9. Krakauer, J.W., Z.M. Pine, M.F. Ghilardi & C. Ghez. 2000. Learning of visuomotor transformations for vectorial planning of reaching trajectories. *J. Neurosci.* **20:** 8916–8924.
10. Kagerer, F.A., J.L. Contreras-Vidal & G.E. Stelmach. 1997. Adaptation to gradual as compared with sudden visuomotor distortions. *Exp. Brain Res.* **115:** 557–561.
11. Hwang, E.J., M.A. Smith & R. Shadmehr. 2006. Dissociable effects of the implicit and explicit memory systems on learning control of reaching. *Exp. Brain Res.* **173:** 425–437. doi: 10.1007/s00221-006-0391-0
12. Saijo, N. & H. Gomi. 2010. Multiple motor learning strategies in visuomotor rotation. *PloS One* **5:** e9399. doi: 10.1371/journal.pone.0009399
13. Fernandez-Ruiz, J., W. Wong, I.T. Armstrong & J.R. Flanagan. 2011. Relation between reaction time and reach errors during visuomotor adaptation. *Behav. Brain Res.* **219:** 8–14. doi:10.1016/j.bbr.2010.11.060
14. Benson, B.L., J.A. Anguera & R.D. Seidler. 2011. A spatial explicit strategy reduces error but interferes with sensorimotor adaptation. *J. Neurophysiol.* **105:** 2843–2851. doi: 10.1152/jn.00002.2011
15. Martin, T.A., J.G. Keating, H.P. Goodkin, *et al.* 1996. Throwing while looking through prisms. II. Specificity and storage of multiple gaze-throw calibrations. *Brain* **119**(Pt 4): 1199–1211.
16. Anguera, J.A., P.A. Reuter-Lorenz, D.T. Willingham & R.D. Seidler. 2010. Contributions of spatial working memory to visuomotor learning. *J. Cogn. Neurosci.* **22:** 1917–1930. doi: 10.1162/jocn.2009.21351
17. Keisler, A. & R. Shadmehr. 2010. A shared resource between declarative memory and motor memory. *J. Neurosci.* **30:** 14817–14823. doi: 10.1523/JNEUROSCI.4160-10.2010
18. Eversheim, U. & O. Bock. 2001. Evidence for processing stages in skill acquisition: a dual-task study. *Learn. Mem.* **8:** 183–189. doi: 10.1101/lm.39301
19. Taylor, J.A. & K.A. Thoroughman. 2008. Motor adaptation scaled by the difficulty of a secondary cognitive task. *PloS One* **3:** e2485. doi: 10.1371/journal.pone.0002485
20. Taylor, J.A. & K.A. Thoroughman. 2007. Divided attention impairs human motor adaptation but not feedback control. *J. Neurophysiol.* **98:** 317–326. doi: 10.1152/jn.01070.2006
21. Galea, J.M., S.A. Sami, N.B. Albert & R.C. Miall. 2010. Secondary tasks impair adaptation to step- and gradual-visual displacements. *Exp. Brain Res.* **202:** 473–484. doi: 10.1007/s00221-010-2158-x
22. Malone, L.A. & A.J. Bastian. 2010. Thinking about walking: effects of conscious correction versus distraction on locomotor adaptation. *J. Neurophysiol.* **103:** 1954–1962. doi: 10.1152/jn.00832.2009
23. Mazzoni, P. & J.W. Krakauer. 2006. An implicit plan overrides an explicit strategy during visuomotor adaptation. *J. Neurosci.* **26:** 3642–3645. doi: 10.1523/JNEUROSCI.5317-05.2006
24. Taylor, J.A. & R.B. Ivry. 2011. Flexible cognitive strategies during motor learning. *PLoS Comput. Biol.* **7:** e1001096. doi: 10.1371/journal.pcbi.1001096
25. Thoroughman, K.A. & R. Shadmehr. 2000. Learning of action through adaptive combination of motor primitives. *Nature* **407:** 742–747. doi: 10.1038/35037588
26. Cheng, S. & P.N. Sabes. 2006. Modeling sensorimotor learning with linear dynamical systems. *Neural Comput.* **18:** 760–793. doi: 10.1162/089976606775774651
27. Zarahn, E., G.D. Weston, J. Liang, *et al.* 2008. Explaining savings for visuomotor adaptation: linear time-invariant state-space models are not sufficient. *J. Neurophysiol.* **100:** 2537–2548. doi: 10.1152/jn.90529.2008.
28. Izawa, J. & R. Shadmehr. 2011. Learning from sensory and reward prediction errors during motor adaptation. *PLoS Comput. Biol.* **7:** e1002012. doi: 10.1371/journal.pcbi.1002012
29. Shabbott, B.A. & R.L. Sainburg. 2010. Learning a visuomotor rotation: simultaneous visual and proprioceptive information is crucial for visuomotor remapping. *Exp. Brain Res.* **203:** 75–87. doi: 10.1007/s00221-010-2209-3
30. Hinder, M.R., J.R. Tresilian, S. Riek & R.G. Carson. 2008. The contribution of visual feedback to visuomotor

adaptation: how much and when? *Brain Res.* **1197:** 123–134. doi: 10.1016/j.brainres.2007.12.067
31. Hinder, M. R., D. G. Woolley, J.R. Tresilian, *et al.* 2008. The efficacy of colour cues in facilitating adaptation to opposing visuomotor rotations. *Exp. Brain Res.* **191:** 143–155. doi: 10.1007/s00221-008-1513-7
32. Huang, V.S., A. Haith, P. Mazzoni & J.W. Krakauer. 2011. Rethinking motor learning and savings in adaptation paradigms: model-free memory for successful actions combines with internal models. *Neuron* **70:** 787–801. doi: 10.1016/j.neuron.2011.04.012.
33. Verstynen, T. & P.N. Sabes. 2011. How each movement changes the next: an experimental and theoretical study of fast adaptive priors in reaching. *J. Neurosci.* **31:** 10050–10059. doi: 10.1523/JNEUROSCI.6525-10.2011
34. Diedrichsen, J., O. White, D. Newman & N. Lally. 2010. Use-dependent and error-based learning of motor behaviors. *J. Neurosci.* **30:** 5159–5166. doi: 10.1523/JNEUROSCI.5406-09.2010
35. Smith, M.A. & R. Shadmehr. 2005. Intact ability to learn internal models of arm dynamics in Huntington's disease but not cerebellar degeneration. *J. Neurophysiol.* **93:** 2809–2821. doi: 10.1152/jn.00943.2004
36. Rabe, K. *et al.* 2009. Adaptation to visuomotor rotation and force field perturbation is correlated to different brain areas in patients with cerebellar degeneration. *J. Neurophysiol.* **101:** 1961–1971. doi: 10.1152/jn.91069.2008
37. Criscimagna-Hemminger, S.E., A.J. Bastian & R. Shadmehr. 2010. Size of error affects cerebellar contributions to motor learning. *J. Neurophysiol.* **103:** 2275–2284. doi: 10.1152/jn.00822.2009
38. Lang, C.E. & A.J. Bastian. 2002. Cerebellar damage impairs automaticity of a recently practiced movement. *J. Neurophysiol.* **87:** 1336–1347.
39. Taylor, J.A., N.M. Klemfuss & R.B. Ivry. 2010. An explicit strategy prevails when the cerebellum fails to compute movement errors. *Cerebellum* **9:** 580–586. doi: 10.1007/s12311-010-0201-x
40. Martin, T.A., J.G. Keating, H.P. Goodkin, *et al.* 1996. Throwing while looking through prisms. I. Focal olivocerebellar lesions impair adaptation. *Brain* **119**(Pt 4): 1183–1198.
41. Ito, M. 2008. Control of mental activities by internal models in the cerebellum. *Nat. Rev. Neurosci.* **9:** 304–313. doi: 10.1038/nrn2332
42. Strick, P.L., R.P. Dum & J.A. Fiez. 2009. Cerebellum and nonmotor function. *Ann. Rev. Neurosci.* **32:** 413–434. doi: 10.1146/annurev.neuro.31.060407.125606
43. Pascual-Leone, A., E.M. Wassermann, J. Grafman & M. Hallett. 1996. The role of the dorsolateral prefrontal cortex in implicit procedural learning. *Exp. Brain Res.* **107:** 479–485.
44. Slachevsky, A. *et al.* 2001. Preserved adjustment but impaired awareness in a sensory-motor conflict following prefrontal lesions. *J. Cogn. Neurosci.* **13:** 332–340.
45. Slachevsky, A. *et al.* 2003. The prefrontal cortex and conscious monitoring of action: an experimental study. *Neuropsychologia* **41:** 655–665.
46. Ivry, R.B., J. Schlerf, J. Xu, *et al.* 2008. Strategic and recalibration processes during visuomotor rotation in cerebellar ataxia. *Soc. Neurosci. Abstr.* Washington, DC.
47. Fernandez-Ruiz, J., C. Hall, P. Vergara & R. Diiaz. 2000. Prism adaptation in normal aging: slower adaptation rate and larger aftereffect. *Brain Res. Cogn. Brain Res.* **9:** 223–226.
48. Heuer, H. & M. Hegele. 2008. Adaptation to visuomotor rotations in younger and older adults. *Psychol. Aging* **23:** 190–202. doi: 10.1037/0882-7974.23.1.190
49. Kounios, J. & M. Beeman. 2009. The Aha! Moment: the cognitive neuroscience of insight. *Curr. Dir. Psychol. Sci.* **18:** 210–216.
50. Bowden, E.M. & M. Jung-Beeman. 2003. Aha! Insight experience correlates with solution activation in the right hemisphere. *Psychon. Bull. Rev.* **10:** 730–737.
51. Smith, R.W. & J. Kounios. 1996. Sudden insight: all-or-none processing revealed by speed-accuracy decomposition. *J. Exp. Psychol. Learn. Mem. Cogn.* **22:** 1443–1462.

Ann. N.Y. Acad. Sci. ISSN 0077-8923

ANNALS OF THE NEW YORK ACADEMY OF SCIENCES
Issue: *The Year in Cognitive Neuroscience*

# Efficient coding and the neural representation of value

Kenway Louie[1] and Paul W. Glimcher[1,2,3]
[1]Center for Neural Science, [2]Department of Psychology, [3]Department of Economics, New York University, New York, New York

Address for correspondence: Kenway Louie, Center for Neural Science, New York University, 4 Washington Place, Room 809, New York, NY 10003. klouie@cns.nyu.edu

**To survive in a dynamic environment, an organism must be able to effectively learn, store, and recall the expected benefits and costs of potential actions. The nature of the valuation and decision processes is thus of fundamental interest to researchers at the intersection of psychology, neuroscience, and economics. Although normative theories of choice have outlined the theoretical structure of these valuations, recent experiments have begun to reveal how value is instantiated in the activity of neurons and neural circuits. Here, we review the various forms of value coding that have been observed in different brain systems and examine the implications of these value representations for both neural circuits and behavior. In particular, we focus on emerging evidence that value coding in a number of brain areas is context dependent, varying as a function of both the current choice set and previously experienced values. Similar contextual modulation occurs widely in the sensory system, and efficient coding principles derived in the sensory domain suggest a new framework for understanding the neural coding of value.**

**Keywords:** decision making; context dependence; reward; efficient coding; neuroeconomics

## Introduction

The activity in many brain regions is sensitive to reward, a modulation that can reflect not just value but processes such as sensation, motivation, and attention. Parallel work on humans and animals has begun to elucidate the different functional roles of areas representing value itself: responses linked to the values of specific actions in decision-related areas, activity that represents and perhaps stores action-independent value in the frontal cortices, and value-related teaching signals in subcortical regions that guide learning. These investigations provide details about the mechanism of value representation, and growing evidence is revealing subtle but important consequences resulting from these implementations of value representation in neural activity. For example, although normative models of choice assume that the values of options or goods are evaluated in an absolute manner, independent of other available alternatives, the neural representation of value has been shown to depend significantly on choice context. This finding not only carries implications for behavior but also suggests that the neural encoding of value reflects key features of efficient encoding systems first identified in the visual and auditory systems. In this review, we consider the different forms of context-dependent value representation observed in the brain and compare them to well-known contextual effects in sensory processing.

We begin by reviewing the different forms of value representation observed in the brain, with a focus on results from primate electrophysiology. We then describe the ways that neural value coding depends on both spatial and temporal value context. In the following section, we describe the different ways that spatial and temporal context affect sensory processing. Finally, we briefly review the tenets of the *efficient coding hypothesis* as it applies to sensory systems and address how these context-dependent value representations can be viewed within the efficient coding hypothesis.

## Value-related activity in the brain

To efficiently interact with its environment, an organism must be able to predict the consequences of actions and choose the best of possible alternative

doi: 10.1111/j.1749-6632.2012.06496.x

options. Value, as a quantification of the expected rewards or costs associated with any choice or action, is thus critical to the decision-making process. This fundamental relationship between value and choice is expressed explicitly in economic theory, which defines the *expected utility* of an object only from an analysis of the choices a decision-maker makes between that object and other options.[1,2] In this regard, economic theories respect the fact that, for example, a given chooser might view 10 apples as less than 10 times as good as one apple if that is what the subject's choices reveal. In neuroscientific studies of reward and decision making, however, the experimental parameter manipulated is typically an objective quantity such as the number of apples or the amount of liquid reward delivered to the animal, and it is this objective quantity that neurobiologists have typically hypothesized is encoded in the nervous system. Motivated by economic models of choice, a growing number of neuroscientific studies have demonstrated that it is in fact the subjective rather than objective value of rewards that best correlates with reward-related activity in the brain (for more information, see Refs. 3 and 5).

For an organism facing an uncertain and dynamic world, however, optimal behavior requires more than just using value to guide decision making: values must be learned through interaction with the environment, and these values must be stored and updated over time for subsequent use. Consistent with this idea, evidence from neurophysiological studies indicate that value-related neural activity is observed widely throughout the brain and there is growing evidence that activity in these value-related areas can be broadly grouped into three categories according to function: action selection, value storage, and learning. In line with the idea that value is an intrinsic part of the decision process, many of the brain systems involved in action selection and decision making are modulated by value. There is accumulating evidence that frontal regions, notably the ventromedial prefrontal and orbitofrontal cortices, encode the value of options independent of the actions required to achieve them; such a representation is consistent with these areas being involved in the storage and recall of values. Finally, neural circuits that respond near the time of reward receipt are systematically modulated by value in a manner consistent with a role in learning. We briefly review each of these in turn, with a focus on the primate electrophysiology literature.

Multiple stages of the action-selection process are significantly modulated by reward, with neural activity covarying with the value of actions in these stages. For example, in the primate visuo-saccadic system, value-coding activity is observed at both cortical and subcortical levels of processing. Early in the sensorimotor processing pathway, the activity of neurons in the posterior parietal cortex varies monotonically with the subjective value of the reward associated with a saccade.[6–10] This influence of value extends through the oculomotor pathway to brain areas more closely tied to saccade execution, with reward expectation modulating activity in the frontal eye fields,[11] supplementary eye fields,[12] and the superior colliculus.[13,14] Reward-related activity is also readily observed in the basal ganglia, where the activity of striatal neurons reflects the expectation of reward.[15–17] When reinforcement learning models are fit to behavioral data in dynamic tasks, many striatal neurons encode the derived trial-by-trial action values.[18,19] In these saccade-related brain areas, neurons retain their spatial selectivity, and the influence of value acts primarily like a modulation of gain. Although most of the accumulated evidence pertains to the oculomotor system, similar valuation signals are likely to exist in other effector systems, such as that controlling reach.[18,20,21]

Although decision making can be viewed as purely a process of action selection, values can also be associated with choice options in a manner independent of motor action. Such a "goods-based" system provides flexibility, allowing an animal to enact a decision independent of simple stimulus–response associations. A recent experiment by Padoa-Schioppa and Assad suggests that neurons in area 13 of the primate orbitofrontal cortex (OFC) can encode exactly this kind of goods-based value.[22] Employing a binary choice task that required animals to choose between varying amounts of two different juice types, the authors determined the relative value of the two juice rewards. In contrast to the decision- and motor-related areas mentioned previously, they found that OFC activity was generally insensitive to the spatial configuration of stimuli or the required motor action. However, many of these neurons showed one or more of three particular responses linked to the choice options: *offer type* responses, which varied with the value (or amount) of one of the offered options; *chosen value* responses, which varied with the value of the selected option; and *taste* responses, which varied in a binary fashion, depending on which juice type was selected. Of these different factors, chosen value most closely resembles an economic, goods-based notion of value; it is a subjective quantity, dependent on an individual animal's valuation of the two juices, and represents value independent of the identity of the particular chosen option.

Unlike the action-value representations appearing in effector-specific circuits, value signals in OFC appear to be present throughout the time course of the choice process, both before and after the receipt of reward, suggesting that they may inform rather than directly participate in action selection. However, such value coding may play a role in economic decision making independent of motor response.[23] This demonstration of a goods-based, subjective value representation aligns with previous neurophysiological reports of OFC responses to reward expectation.[24–27] In human neuroimaging studies, a growing number of studies have reported activity related to reward expectation and subjective value in the ventromedial prefrontal cortex (vmPFC; see Kable and Glimcher[4] and Rangel and Hare[5] for recent reviews).[4,5] Thus, it appears that frontal circuits in humans convey a unitary representation of action-independent reward value as well. Similar action-independent value representations may occur in related brain regions, as both the OFC and the vmPFC are heavily interconnected with structures implicated in reward processing, such as the cingulate cortex, the amygdala, and the hippocampus;[28] many neurons in the amygdala, for example, encode both the positive and negative values of conditioned stimuli or states.[29,30]

Other brain areas show reward-related activity that is more consistent with a role in learning rather than in the expression or storage of value. Such a system is critical because animals must, in the real world, update the value of choice options through continuing experience. Neurons in the midbrain dopaminergic system are now known to encode a teaching signal well-suited to updating stored value estimates. These neurons influence a large number of brain structures involved in motivation and goal-directed behavior, with projections targeting the nucleus accumbens and frontal cortex via the mesolimbic and mesocortical pathways and the striatum via the nigrostriatal pathway. Multiple lines of evidence have long suggested that dopamine is involved in the processing of reward: drugs of abuse, such as cocaine, nicotine, and amphetamines, indirectly or directly increase the action of dopamine, and dopaminergic pathways are among the most effective locations for the placement of intracranial electrical self-stimulation electrodes.[31,32]

Two developments in the 1990s clarified the role of dopamine in reward processing in a fundamental manner. First, in a series of electrophysiological studies of primates, Schultz and colleagues demonstrated that dopamine neurons show a phasic response to appetitive rewards, such as a small piece of apple or quantity of juice.[33–35] Significantly, when visual stimuli are associated with rewards through a classical conditioning paradigm, the phasic dopaminergic response to reward delivery diminishes; instead, dopamine neurons respond at the presentation of the conditioned, predictive stimulus. This transition of neural response from the reward delivery to the conditioned stimulus mirrors the transfer of the animals' behavioral reactions. Importantly, these responses also carry information about the timing of expected rewards; when an expected reward delivery is omitted, dopaminergic neurons show a marked suppression of activity at the time when reward would have occurred.[36]

In the second of these critical advances, Montague and colleagues used computational learning theory to provide a theoretical framework for understanding these dopaminergic neuronal responses[37,38] (for a more complete review of these advances, see Glimcher[39]). Learning based on prediction errors, in which an organism makes a prediction and learns contingent on errors in that prediction, is a process central to adaptation and learning rules in both computer science and psychology.[40,41] Building on earlier studies of octopaminergic neurons in the honeybee brain, these authors suggested that mesencephalic dopamine neurons encode an error prediction that provides a dynamic signal of the difference between the expected amount of reward and the actual reward. Montague and colleagues proposed that this dopaminergic reward prediction error (RPE) signal drives learning via a temporal difference (TD) algorithm, a type of method in reinforcement learning first introduced to computer science by Sutton and Barto.[42,43] In TD learning, the learned *value function* represents, essentially, the (time-discounted) sum of all expected future rewards associated with any given set of actions. In each iterative learning experience, this value function is updated by a quantity determined by the RPE and the *learning rate* (a model parameter that adjusts the strength with which unexpected rewards update the estimates of future expected rewards). Examination of the primate electrophysiological results in the context of reinforcement learning suggested that dopaminergic responses display the fundamental quantitative and qualitative characteristics of this theoretical RPE signal.[44–46] Thus, unlike the parietal or orbitofrontal responses described previously, dopaminergic responses are related to value but do not appear to encode value itself, but rather a teaching signal used to update value representations elsewhere in the brain.

To review, we have outlined a broad framework loosely categorizing value-related neural activity into action selection, value representation, and value updating. This conceptual organization provides a useful heuristic with which to systematically consider the various functional roles of value-related activity and contextual modulation that we discuss later. However, it is important to note that this heuristic can serve only as a very general guide, and some experimental findings do not fit neatly within this simple framework. One important issue that we have oversimplified concerns the relative roles of action- and goods-based selection and their integration in decision making. One possibility that has been raised by a number of authors is that all decisions are implemented at the stage of action selection, when information about option value (also called goods or stimulus value) is combined with information about motor costs into action values to guide choice.[5] Alternatively, it has been suggested that many or all decisions may be made in an abstract goods space independent of motor implementation and occur solely in brain regions such as OFC.[23] Finally, goods-based and action-based representations may coexist and decisions might be accomplished in either framework depending on the nature of any given task.[47] A better understanding of the precise mechanism of valuation and decision making in these kinds of choices will be critical for predicting the effects of the different kinds of contextual modulation observed in these areas and discussed later.

Another important area of simplification concerns the fact that many of these brain regions may play a role in other functional processes, and these alternative signals may confound our interpretation of the role of these areas in valuation and decision making. For example, value itself often co-varies with other important behavioral quantities, such as motivation and attention.[48] This is a particularly relevant issue in the parietal cortex, where lateral intraparietal area (LIP) neurons are known to be modulated by visual salience and the allocation of spatial attention.[49,50] Such findings have led some to suggest that activity in the parietal cortex reflects the allocation of spatial attention and plays little or no direct role in decision making.[51] Given the tight behavioral correlation between attention and decisions about eye and arm movements, we take a middle ground in this review, suggesting that both quantities are represented. Indeed, the observation that attention and decision are for the most part behaviorally inseparable has led us to suggest elsewhere that it is unlikely that these two quantities are fully orthogonalized in the nervous system.

Finally, we note that some brain areas may be involved in more than one of these stages in the processing of value information. For example, in an oculomotor foraging task, some neurons in the striatum are correlated with the value of a specific directional saccade (action value), but others are instead correlated with the value of the saccade that is chosen, regardless of direction.[18,19] These chosen value neurons cannot guide selection: their activity is contingent on the selected action, and they are most active immediately before or persistently after reward delivery. However, because such activity represents the expected outcome of any action, it may play a role in updating stored representations of value (for example, the difference between chosen value and obtained reward can be used as a teaching signal).

## Contextual modulation in value representation

In normative models of decision making, such as those in neoclassical economics, a fundamental assumption is that options are evaluated in an absolute manner, and that the values assigned to goods or actions are stable, stationary quantities.[52,53] Under this kind of value representation, a decision-maker will have a complete preference order over all possible choice options, and will always choose the highest valued option from a set of possible alternatives. However, growing behavioral evidence indicates that choice behavior in both animals and humans is often context dependent, for example, varying depending on the composition of the choice set or exposure to previous priming situations. Choice behavior is also to a certain degree stochastic, varying somewhat randomly from moment to moment and trial to trial. These behavioral data suggest that choice processes rely on a noisy and comparative form of evaluation, driven by a relative representation of value dependent on both spatial and temporal context. Below, we review the emerging neurophysiological evidence for context-dependent effects on the activity of value coding areas.

### Spatial context and value coding

Many computational theories of action selection and decision making require a representation of the value of individual actions, from which a single action is then selected. However, the precise relationship between value and neural firing rates is not known. Although the concept of utility in economic models of choice is unattached to a particular unit of measure (that is, utility is *ordinal* not *cardinal and unique*), the neural representation of value is instantiated via actual spiking rates, which are necessarily fully cardinal unique values. As a result, many different possible neural representations of value will be consistent with a given set of choice data (and a given ordinal ranking); for example, systems whose value representations are linear transforms of one another (e.g., $V_1 = 10$, $V_2 = 20$, and $V_3 = 30$ spikes/sec vs. $V_1 = 50$, $V_2 = 100$, and $V_3 = 150$ spikes/sec) would produce identical behavioral choice preferences. Thus, behaviorally generated models of value only provide limited constraints on how neural systems represent values.

Many economic models thus do not distinguish between representations coding absolute value, where action values are modulated strictly by the value of the target option, and those coding relative value, in which action values are normalized to the value of all available options. In the animal decision-making literature, however, there is a strong historical precedent for the idea that value is represented in a fractional manner:

$$FV_1 = \frac{V_1}{V_1 + V_2},$$

where $FV_1$ is the fractional value of option 1, and $V_1$ and $V_2$ are the values of the two options in the choice set. In studies of foraging behavior in pigeons, Herrnstein established that the relative response probability for a given option was intricately tied to the relative rate of reinforcement.[54] This became the basis of the influential *matching law*, which proposes that the fraction of choices an animal allocates to a given option will match the fraction of rewards earned from that option. Under this proposal, the primary determinant of choice behavior is the relative (fractional) value of rewards.

Does the brain represent action values in absolute terms, independent of the other available options, or in relative terms? In visuomotor areas like the parietal cortex, neurons display visual and motor selectivity that coincides in space, suggesting that they link sensory and motor information during decision making. Thus, a given neuron in LIP represents a specific action—a saccadic eye movement to its response field (RF). In different paradigms, decision-related LIP activity is modulated by a number of behaviorally relevant factors that affect the choice of saccade, including accumulated motion evidence, target color, temporal information, and probabilistic cues.[55–58] One unifying hypothesis for these various correlates is that they control LIP activity by influencing the subjective value of the associated saccades. The representation of the value of a saccade (action value) is seen as a modulation of the action-selective activity, a finding now demonstrated in a number of studies.[6–8,10,59] Growing experimental evidence suggests that these value signals reflect the relative rather than absolute value of a given saccade. LIP activity correlates with the value of the response field target when that quantity is varied alone. However, if the value of both targets in a two-target task are varied, LIP responses depend on both the response field target value ($V_{in}$) and the extra-response field target value ($V_{out}$), consistent with a fractional, relative reward representation.[7,59] For example, Rorie *et al.* recorded LIP neurons while monkeys performed a classic perceptual discrimination task, which requires animals to judge the motion direction of a noisy motion-dot stimulus.[59] LIP neurons display characteristic decision-related activity in this paradigm,[55,60] with firing rate increases paralleling the accumulating sensory evidence for a given saccade. When Rorie *et al.* manipulated the values associated with the two choices, they found that LIP activity reflected the absolute value of the RF target, consistent with previous studies. More significantly, they found that LIP neurons also reflected the *relative* value of the RF target, with higher firing rates for a given RF target reward if the other, extra-RF target was associated with a low-value reward.

How is a relative value representation constructed in decision circuits? If relative value representation is considered in terms of spatial context, activity driven by the response field target value is suppressed by the value of other targets situated outside the RF. In sensory cortices, stimuli outside the classical receptive field can nonetheless significantly modulate neuronal activity driven by receptive field stimulation. Many of these extra-classical effects are characterized by models in which response is specified by the sensory properties of the stimulus inside the receptive field, divided by the weighted sum of the sensory properties of stimuli both outside and inside the receptive field.[61] For example, the output of a cell in the visual cortex can be described as

$$R_i \propto \frac{A_i}{\sigma^2 + \sum_j A_j},$$

where $A_i$ is the driving input of the cell in question, the summation in the denominator is taken over inputs $A_j$ to a large population of similar neurons, and $\sigma^2$ is an empirical semisaturation constant. This *divisive normalization* mechanism is widely found in the visual cortex and explains nonlinear phenomena in the striate and extra-striate cortex[62–65] as well as object-driven normalization in the ventral visual stream.[66] Furthermore, divisive normalization produces an efficient coding of natural signals[67,68] and may underlie the attentional modulation of neural responses,[69] suggesting that it may be a canonical computational algorithm in the cortex.[70] If the parietal cortex employs an analogous functional architecture, then a similar form of divisive normalization may underlie the representation of saccade value in the LIP.

In a recent study, we explicitly examined the influence of alternative option values on LIP activity.[71] To confirm a relative rather than absolute coding of value, we first quantified LIP responses in a two-target task, in which the RF target value was held constant and the extra-RF target value was explicitly varied. Consistent with other reports,[7,59] we found that the activity of both single neurons and the population is inversely related to the value of the alternative target, suggesting that action values are coded relative to the choice context. To more precisely quantify the nature of this modulation, we recorded additional LIP neurons in a three-target task in which we systematically varied the number of saccade targets and their values, enabling us to test the divisive normalization model. Monkeys fixated a central cue and were presented with either one, two, or three targets, each of which was associated with a different magnitude of water reward. After target presentation, monkeys were subsequently instructed to select one of the presented targets: a medium reward target situated in the RF or either the small or large reward targets placed outside the RF, typically in the opposite hemifield. Each trial consisted of one of seven possible target arrays, presented randomly and with equal probability (three single target, three dual target, and one triple target trial). Each randomized target array provided a unique combination of value associated with the target in the RF and values available outside the RF (Fig. 1A), allowing us to quantify the relationship between target value ($V_{\text{in}}$) and value context ($V_{\text{out}}$).

We found two primary effects of the value context, defined by the instantaneous choice set, on LIP responses (Fig. 1B). First, consistent with qualitative reports in two target tasks, activity elicited by target onset in the RF is modulated by the value of the alternatives, with larger $V_{\text{out}}$ magnitudes leading to greater suppression. Second, activity when no RF target is present is suppressed in a context dependent manner, with larger $V_{\text{out}}$ values driving activity further below baseline activity levels. Analogous to extra-classical modulation in the early visual cortex, both of these effects are driven by the value of targets that themselves do not drive the recorded neuron. Significantly, when we performed a quantitative model comparison with other possible relative reward representations including the fractional value of Herrnstein, $V_{\text{in}}/(V_{\text{in}}+V_{\text{out}})$, and a differential value model, $V_{\text{in}}-V_{\text{out}}$, contextual value modulation was clearly best explained by a divisive normalization-based model:

$$R_i \propto \frac{V_i + \beta}{\sigma^2 + \sum_j V_j},$$

where the activity of a neuron $R_i$ is dependent on both the value of the target in its RF $V_i$ and the sum over all available target values $V_j$ (the empirical parameter $\beta$ models suppression below

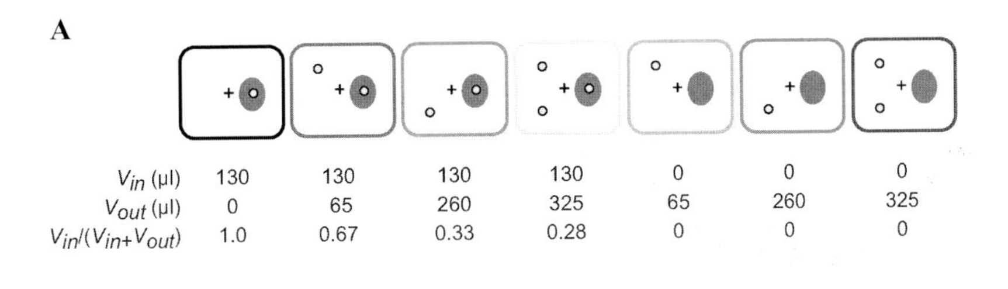


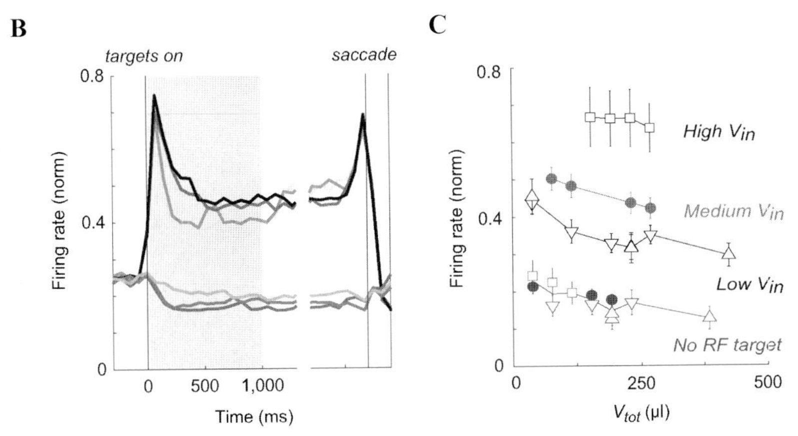


**Figure 1.** Spatial context dependence in LIP value coding. (A) Different value conditions in an oculomotor saccade task. Monkeys were presented with a target array of one, two, or three peripheral targets associated with different reward magnitudes. The value of the RF target was constant, whereas the value context varied with the number and reward magnitude of extra-RF targets. (B) Population parietal neuron activity. The value context (extra-RF target value) modulates LIP activity both in the presence and absence of the target in the RF. (C) Neural coding of value and value context. Increasing the value of the RF target increases LIP activity (different lines). Increasing the value of extra-RF targets suppresses LIP activity (connected lines). Together, these data suggest that LIP activity encodes a relative reward representation incorporating both target value and value context. (Adapted from Louie *et al.*[71])

a baseline rate). Significantly, the implementation of relative value through divisive normalization suggests a functional linkage to contextual modulation in sensory systems, which use an analogous normalization algorithm over sensory inputs.

### *Temporal context and value coding*

A fundamental question about the neural representation of value is how such coding changes with behavioral context; in other words, is value coding relative or absolute? The results described previously suggest that at least some parietal circuits involved in decision making reflect a normalization process across the available choice options, but it is not yet entirely clear how this parietal circuit fits into the larger network of areas involved in decision making. A number of brain regions in addition to the parietal cortex display decision-related activity during the choice process, such as the basal ganglia, frontal eye fields, supplementary eye fields, and superior colliculus. A remaining open question is whether normalization is a general aspect of value coding at all stages of action selection. We do know, however, that neural activity in the monkey dorsal premotor cortex during reach decisions is modulated by the relative value of reach targets,[72] indicating that value normalization may be a general phenomenon. Thus, one important issue is how contextual value coding varies in different brain areas performing different value-related processing. It is reasonable to expect that action value–based gain control, presumably occurring in an instantaneous manner when the options are presented, may be a feature specific to areas involved in decision making, and that value-sensitive brain regions subserving other functions—such as value storage—employ different value coding strategies.

Given the demonstration that orbitofrontal neurons encode a goods-based representation of value, Padoa-Schioppa and Assad examined whether those value representations are dependent on the other available rewards in a choice situation.[73] As in their original demonstration of value coding by OFC neurons,[22] monkeys chose between varying amounts pairs of juices (A:B, B:C, C:A) that could be ranked by relative preference order (when offered in equal amounts, A > B > C). Trials with different pairwise reward offers were randomly interleaved. Monkeys in this task displayed transitivity, for example, choosing 1A over 1C if they chose 1A over 1B and 1B over 1C, indicating that the different rewards could be compared on a common value scale, enabling the examination and comparison of the different neural value representations. As before, the authors found three general types of response, which they termed *offer value* (the presented value of a specific reward type), *chosen value* (the value of the selected option in a given trial, regardless of type), and *taste* (received reward type). When the value-specific responses (offer and choice value) were examined, they did not depend on the specific pair of rewards offered: for example, the activity of a neuron encoding *offer value A* had the same linear relationship between firing rate and offered amount of A whether the other available reward was B or C. The authors concluded that OFC responses are invariant to the menu of choice options, and do not reflect the relative preference ranking of the possible rewards.

This menu invariance appears at first glance to contradict earlier results by Tremblay and Schultz, describing relative reward preference in OFC neurons.[25] In that study, monkeys performed a spatial delayed-response task in which they were presented with a stimulus that predicted which of two possible rewards would be delivered at the end of the trial; a single trial consisted of one stimulus that was associated with a specific liquid or food reward. The task was conducted in blocks of trials, with two different stimuli and their respective associated rewards employed in a given block. Of the OFC neurons active in this task, many showed reward-related responses, responding in a phasic manner to the instruction stimulus or reward delivery and in a sustained activation preceding reward. These responses often showed greater activation for one kind of reward over others but did not differentiate between left and right instructions or different cues indicating the same reward, suggesting that they reflected information about the predicted reward.

Notably, in contrast to the Padoa-Schioppa and Assad findings, OFC neurons in the Tremblay and Schultz task encoded a value representation that was relative rather than absolute. In separate choice trials, the authors established the relative behavioral preferences between each of three different reward pairs (A:B, B:C, and C:A). The majority (40 of 65) of reward sensitive neurons showed reward responses that were dependent on the block context, with different activation for a given predicted reward contingent on which other reward was available in the block. For example, in a monkey that preferred reward A to reward B and reward B to reward C, this kind of response would show low activation in a B trial when both A and B were offered in a block but high activation in a B trial when B and C were paired. Note that these differential responses occurred in trials with identical visual stimuli and rewards; only the larger context of which reward was available in other trials varied. These results suggest that OFC neurons encode a subjective, context-dependent value driven by relative preference rather than absolute, unchanging properties of the reward itself.

How can menu invariance and relative reward preferences both occur in orbitofrontal neurons? A distinct difference between the two experiments is that the different pairwise combinations of rewards were presented randomly interleaved in the study reporting menu invariance, whereas the study showing relative reward preference presented reward pairs in blocks. If the orbitofrontal value representation adapts to the recent history of received rewards, relative reward coding over time would appear as differential adaptation occurs to the rewards in different blocks of contiguous trials. When the choice context changes rapidly with randomly interleaved pairwise rewards, the local distribution will appear almost identical to the global distribution if the integration time for adaptation is sufficiently larger, and value representations will appear absolute, or invariant.

Additional work by Padoa-Schioppa indicates that the results described in the two preceding studies can indeed be reconciled by accounting for the temporal dynamics of how different reward possibilities were presented.[74] This study consisted primarily of a reanalysis of two large datasets presented previously, including data from the study demonstrating menu invariance. In these experiments, animals chose between different amounts of two (or in some cases, three) types of juice rewards. The distribution of possible reward sizes for a given juice type were fixed for each neuron, but varied across neurons. For example, one neuron may have been recorded with B rewards ranging from 0 to 2 (in equivalent units of juice A, determined by behavior), whereas a separate neuron was recorded with B rewards ranging from 0 to 10. To examine value-based adaptation, the authors examined whether, across the population of OFC neurons, firing rates depended on the value range (an example of local value distribution).

One straightforward model of range adaptation is that the firing rate range is adjusted to match the range of possible values. Under this model, the slope of the relationship between firing rate and value would decrease as the possible value range increases, and the high end of the value range should be represented by the same firing rate in different value-range conditions (Fig. 2, top). When the mean population firing rates were examined in this manner, segregated by value range, OFC activity showed a clear adaptation to the locally experienced range of values, for both offer value and chosen value responses (Fig. 2, bottom). Consistent results were observed when individual neurons were recorded under both low-range and high-range conditions, indicating that range adaptation is not an artifact of averaging across the population. Thus, value representation in OFC is independent of the immediate context (menu invariant) but dependent on the local temporal context (range adapting), in contrast to the trial-by-trial normalization observed in areas representing action value.

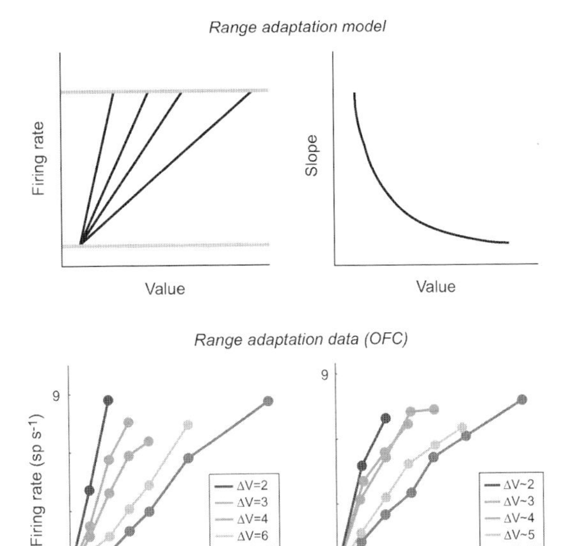


**Figure 2.** Temporal context dependence in OFC value coding. Top: Simple model of range adaptation in value coding neurons. The key assumption is that the range of neural activity is constant across different behavioral value conditions. Bottom: Range adaptation in orbitofrontal neuron activity. The two panels show average OFC activity in two different types of value-coding neurons, color-coded by the range of experienced values (plotted as normalized unit value). Population OFC activity adapts to the range of possible values, indicating that such activity is sensitive to the temporal value context. (Adapted from Padoa-Schioppa.[74])

Although this temporal adaptation can be framed in terms of the range of available values, there are many characteristics that describe the local temporal distribution; for example, in the study described previously, the maximum, mean, and standard deviation of the value distribution varied along with the range. In a recent experiment, Kobayashi *et al.* examined how orbitofrontal neurons adapt their firing rates to reward distributions with different standard deviations but identical means.[75] When individual neurons were exposed to three possible liquid rewards with either a narrow distribution (low standard deviation) or a wide distribution (high standard deviation) of volumes, approximately a quarter of the neurons displayed adaptive coding, with steeper response slopes to the narrow range of rewards, an effect that was significant in the population response. This adaptive coding is of particular interest because it allows the full dynamic range of neural responses to be employed in representing both the narrow and wide distributions. In the light of these effects of value range and standard deviation, the results of Tremblay and Schultz can probably best be interpreted as an adaptation of orbitofrontal firing rates to the mean reward values available in different trial blocks. Importantly, this finding is compatible with the divisive normalization algorithm described previously and observed throughout the cortex, with normalization occurring across time rather than space. These results reinforce the idea that value coding can be adapted to multiple aspects of the (temporally) local probability distribution of values, though the exact parameters of the value distribution that influence adaptation requires further study (e.g., by exploring the effects of higher-order moments like skew).

This kind of adaptive response to the local value statistics may generalize to regions other than the cortex. In a recent study of the dopamine system, widely believed to carry an error prediction signal for updating value representations, Tobler *et al.* examined how recent rewards affect the activity of midbrain dopamine neurons.[76] Monkeys were presented with three different conditioned stimuli, each of which predicted one of two possible rewards that occurred with equal probability. Consistent with coding the difference between outcome and the reward predicted by the cue, dopaminergic activity at the time of reward receipt always increased with the larger reward and decreased with the smaller reward. However, rather than a strict prediction error representing the absolute difference between predicted and received reward, dopamine responses adapted such that better rewards always elicited the same increase in activity regardless of the absolute prediction error. This again represents a form of gain control where the sensitivity of dopaminergic responses is adjusted to the range of possible prediction errors.

### Functional implications of contextual coding

The results reviewed previously suggest that contextual modulation plays an important role in determining the neural coding of value in multiple brain circuits. Although research is just beginning to document such effects, it appears that the nature and extent of contextual influence in a given brain area may be closely tied to its functional role in valuation and decision making. Consistent with a role in storing value information independent of action selection, OFC value coding adapts to the temporal context but appears independent of spatial context. In contrast, consistent with a role in action selection, LIP neurons are strongly influenced by the spatial context of the available choice alternatives. One important open question is how these different forms of contextual modulation are combined in the decision process. For example, the role of the temporal context in decision circuits like the parietal cortex remains unknown; action selection areas may inherit temporally adapted value signals from the frontal cortex, receive value information from nonadapting brain areas and encode value independent of the temporal context, or apply a different temporal weighting function to such nonadapted value signals. Future experimental work will be necessary to fully explore the nature of these different spatial and temporal context effects.

One important and largely unexplored question is the functional consequence of contextual value coding. According to normative, rational theories of choice—such as those in economics—decisions between any pair of options should be independent of the context in which the choice is made.[53,77] For example, the relative preference between any two options should be independent of the presence or value of other alternatives, a property known as *independence from irrelevant alternatives*.[52] However, a large body of behavioral evidence indicates that both animal and human choosers are sensitive to both spatial and temporal forms of context. In trinary choice studies, adding a third low-valued option changes the relative preference between two high-valued options in species ranging from insects to birds to humans.[78–81] Experiments have documented a number of such phenomena in humans, which rely on the alternatives differing in two attribute dimensions, with the effect dependent on the relationship between alternatives in two-dimensional attribute space. A related effect is evident in the so-called *paradox of choice*: despite the rational prediction that more options increase welfare, choosers facing larger choice sets are more likely to select the default option, defer choosing, experience regret, and exhibit inconsistent choice behavior.[82–84]

These behavioral context effects suggest that biological decision making employs some form of comparative valuation, but the mechanism underlying such phenomena remains unknown. One might well hypothesize that relative value coding provides a possible link between decision-making circuits and context-dependent valuation. Contextual modulation, for example normalization in the spatial domain, can significantly alter the relative distance between the mean firing rates that represent different actions. Consider a chooser selecting from three options, two high-value target items and a low-valued distracter item. Under a relative value coding system, like that described previously in the parietal cortex, the mean firing rates representing the values of each option will be divisively scaled by the total value of all alternatives. Accordingly, higher-valued distracters will decrease the distance between the neural representations coding for the values of the target items. Noise, or variability, will critically influence these representations and the choices they produce in such systems, driving increasingly stochastic choice behavior as cardinal representations (neural firing rates) are affected by context. Importantly, variability is an inherent feature of spiking neuron activity: at fixed levels of input, neurons generate action potentials in a stochastic manner, which leads to well-characterized variability in spike counts in repeated measurements.[85–87] Given the increasing evidence for both spatial and temporal forms of contextual modulation in value coding, understanding the interaction between noisy neural systems, context dependence, and stochastic choice behavior remains a key area of future research.

## Contextual modulation in sensory coding

As described previously, there is increasing evidence that the nature of value representation in the brain is dependent on both spatial and temporal context. These responses suggest that the neural mechanisms of value coding are more complex than the classical, behavioral concept of value derived from theories of choice like economics and decision theory. To gain a better understanding of how and why such contextual effects may arise, we turn next to the electrophysiological study of sensory systems, focusing on the visual system where there is well-established literature on the effects of context on both perception and electrophysiological responses.[88] These spatial and temporal context effects are closely linked to the statistics of the natural environment, indicating that sensory circuits are tailored to the signals they are likely to encounter and that contextual modulation may play an important role in information processing.

### *Spatial context in visual coding*

At the psychological level, spatial context produces robust and well-documented effects on perception. Figure 3A illustrates simultaneous contrast, one of the simplest and most powerful examples of spatial context. The two smaller squares are colored the same shade of gray and have the same *luminance* (actual amount of light traveling from the object to the eye), but because of the effect of the surrounding areas the square with the dark surround imparts a higher level of *brightness* (perceived luminance). Spatial context can strongly affect the perception of a wide number of visual features, including brightness, motion, orientation, and object recognition, and may underlie high-level processes like perceptual filling in, figure-ground segregation, and contour detection.[89] Figure 3B shows the tilt illusion, which represents a spatial contextual effect on orientation perception. Here, the surrounding bars oriented 15° counterclockwise away from vertical drives the perceived orientation of the center patch of vertical bars clockwise relative to the true vertical orientation.

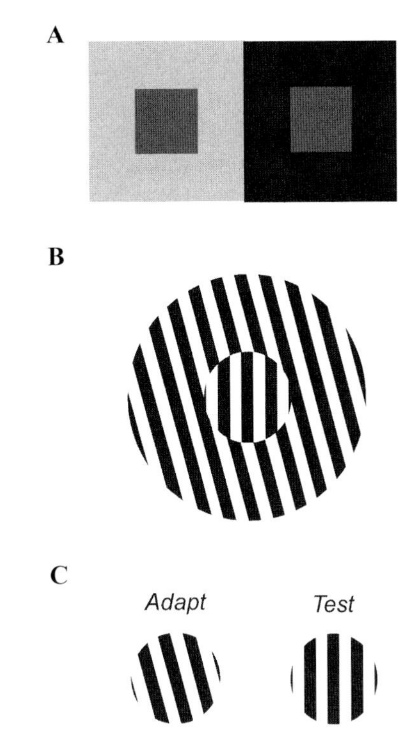


**Figure 3.** Spatial and temporal context effects in visual perception. (A) Simultaneous contrast. The two small squares are an identical shade of gray, but the surrounding spatial contexts drive a differential perception of brightness. (B) Tilt illusion. The presence of a spatial context tilted 15 degrees counterclockwise drives a perceived orientation of the center region that is tilted clockwise from the true vertical orientation. (C) Tilt after-effect. To induce this effect, fixate the adaptation stimulus for 30 sec and then shift fixation to the test stimulus. As in the tilt illusion, the presence of the temporal context (counterclockwise adaptation stimulus) induces a perceived orientation that is repulsed from the context (shifted clockwise).

Do the perceptual effects of context reflect a contextual modulation of neural responses? Neurons in most parts of the visual system respond to visual stimuli in a restricted portion of visual space termed the *classical receptive field* (cRF), a feature that applies to neurons from the earliest stage of visual processing in the retina up to high-level visual cortical areas. This area is typically defined as the portion of the visual field in which a stimulus can elicit spiking activity; by definition, stimuli outside this area do not elicit a spiking response. However, there is a large body of evidence that neurons are modulated by stimuli falling outside the boundaries of the classical receptive field, in an area called the *extra-classical receptive field* (eRF) or *surround*. Many neurons show a differential response to the combination of a stimulus in the cRF and a stimulus in the surround compared to the cRF stimulus alone, an effect that is typically suppressive but can include facilitation as well.

Such contextual interactions are widespread in the visual pathway, and extra-classical effects on neural responses are observed at multiple levels of processing, from retina to lateral geniculate nucleus (LGN) to visual cortical areas, suggesting that contextual modulation may be a fundamental feature of sensory processing. Extra-classical modulation was reported by Hubel and Wiesel in their intial pioneering description of the primary visual cortex (V1), in which they found certain cells (which they called hypercomplex) that were tuned to the length of a bar stimulus, with firing rate increasing with length up to a certain magnitude but attenuating to longer bars. This *end-inhibition* is now known to be one of multiple examples of suppressive modulation driven by stimuli outside the cRF.[90] For example, many V1 neurons show selectivity for the orientation of stimuli within their receptive fields, with a unimodal tuning curve peaked at the optimal stimulus orientation. However, these neurons exhibit a strong contextual modulation termed *iso-orientation suppression*, in which cRF activity is most strongly suppressed by surround stimuli of the same orientation that optimally drives the cRF. Spatial context effects also affect higher levels of visual processing, such as the motion-sensitive neurons of the macaque middle temporal area (area MT). These extrastriate neurons are selective for the direction and speed of motion stimuli in their cRF, with unimodal direction tuning curves similar to V1 selectivity for orientation. Analogous to iso-orientation suppression in V1, stimuli in the surround drive a direction-selective modulation, producing a marked suppression when motion in the surround is the same as motion in the center.

Although spatial contextual modulation appears to be a fundamental feature of visual processing, the anatomic basis and influence on perception of such processes is still an area of active investigation. The mechanisms underlying these spatial modulatory effects may be diverse: although lateral inhibition mediated by intra-area horizontal connections is the standard explanation for surround suppression in V1,[91] feedback projections from higher cortical areas[92] or feedforward inheritance of surround suppression from the LGN[93] may also play a role. There is also a diversity of surround interactions, which can occur with either fast or slower dynamics and drive selectivity changes such as changes in tuning-curve width or shifts in the preferred direction. However, regardless of the underlying mechanism, sensory processing appears to be organized in a manner in which the neural representation of a given feature is a function of the spatial context in which it appears. The ubiquity of center-surround organization and contextual interactions may be related to the inherent structure present in the environment, as this form of processing is well suited to efficiently code natural stimuli[94] (see section "Efficient coding in sensory systems"). The influence of space represents an interaction between the neural representations of different inputs that appear simultaneously, and provides a framework to understand how decision areas encode the value-related activity of multiple choice options, which we explore later.

### *Temporal context in visual coding*

The environment is dynamic, and one of the critical problems faced by the sensory system is processing a constant stream of changing stimuli. The effects of spatial context described previously reflect the interaction of different stimulus features at an instant in time, but stimuli also have a temporal context—the input stimuli in the recent past. The effect of temporal context is referred to by the general term *adaptation*, which describes the response to a sustained presentation of a stimulus (or stimulus distribution).

Like spatial contextual modulation, adaptation is a well-described phenomenon in both the perception and neurophysiology literature. An everyday example of adaptation is the ability to see at different levels of illumination, which is driven by adaptation to ambient illumination in the retina.[95] As an observer moves from a high to a low illumination environment, for example carrying a newspaper from the sunlit outdoors into a darkened room, the perceived brightness of both the dark letters and the gray background remains stable. This adaptation to the local luminance allows the visual system to function over the vast range of possible light levels in the world despite the limited dynamic range of neural firing rates.

The visual system also adapts to a number of higher order visual features beyond simple luminance, though the link between such effects and functional benefits are not as obvious as that for luminance adaptation. There are clear perceptual adaptation effects to features including contrast (relative illumination), orientation, motion, spatial frequency, and even complex objects like faces (for a recent review, see Kohn[96]). Figure 3C illustrates the tilt after-effect, a prominent example of orientation adaptation. Fixating the counterclockwise adaptation stimulus on the left for 30 sec and then shifting fixation to the target stimulus on the right induces a perception that the vertically oriented bars occur tilted clockwise, rotated away from the adapting orientation (a repulsive shift). Interestingly, the tilt after-effect provides a temporal counterpart to the spatial tilt illusion discussed previously; in the tilt illusion, the target and context occur at the same time but separated in space, whereas here the stimuli occur colocalized in space but separated in time.

At the mechanistic level, visual adaptation produces a diverse array of changes in the response of neurons in the visual system. Many of the earliest studies on adaptation studied the effects of presenting two different levels of a stimulus feature, a paradigm that investigates adaptation to the mean of the stimulus distribution. When V1 neurons are exposed to different levels of ambient contrast, the responses shift to encode higher levels of contrast, indicating a decrease in sensitivity.[97] Adaptation to higher-contrast stimuli induces larger reductions in sensitivity, maintaining the neuronal dynamic range close to the average of the recently experienced average contrast. Similar to spatial iso-orientation suppression, suppression of V1 responses is stronger if adapted to stimuli in the preferred versus the opposite or orthogonal orientation.[96] Adaptation to mean responses suppresses activity in a number of visual areas, including MT, V4, and the inferotemporal cortex.

However, when considering the temporal context, the average value of a stimulus is only one way of characterizing the distribution of recent stimuli. A system that responds to the local context of a dynamic environment will be influenced by the shape of the stimulus distribution if the timescale of integration is short relative to the timescale of changes. A number of studies have shown that visual areas can adapt to higher order statistics of local stimulus distributions, such as the variation in stimulus feature. In a study of vertebrate retinal ganglion cells, Smirnakis *et al.* presented random stimuli drawn from distributions with the same mean intensity but differing variances.[98] They found that retinal neurons adapted their responses to the width of the intensity distributions, an effect driven by recent sensory experience. When this temporal modulation is examined in the form of a linear kernel, which shows the average effect of a stimulus as a function of time from presentation, there is a clear temporally weighted dependency on recent intensity. Similar adaptations to stimulus variance have been demonstrated in other visual areas, such as LGN[99] and V1,[100] as well as in other modalities including somatosensation in rats[101] and audition in songbirds.[102] As in spatial contextual modulation, the effects of adaptation are likely mediated by heterogenous mechanisms that differ by both locale of neuronal adaptation and timescale. Nevertheless, adaptation is a widespread feature of sensory processing, suggesting that neural circuits have evolved to respond to both the spatial and temporal statistics of the environment.

## Efficient coding in sensory systems

Given the vast number of possible environmental inputs and the finite amount of neural hardware and metabolic energy possibly devoted to sensory processing, it is natural to assume that sensory systems evolved toward functioning as efficiently as possible. Although there are many possible definitions of efficiency, one enduring and influential proposal for a general principle of sensory system function is the *efficient coding hypothesis*. The fundamental idea is that sensory systems adapt their responses to the regularities of their input, and employ knowledge about these regularities to increase the amount of transmitted information at any given time. This approach relies heavily on the work of Shannon, who developed a quantitative theory of

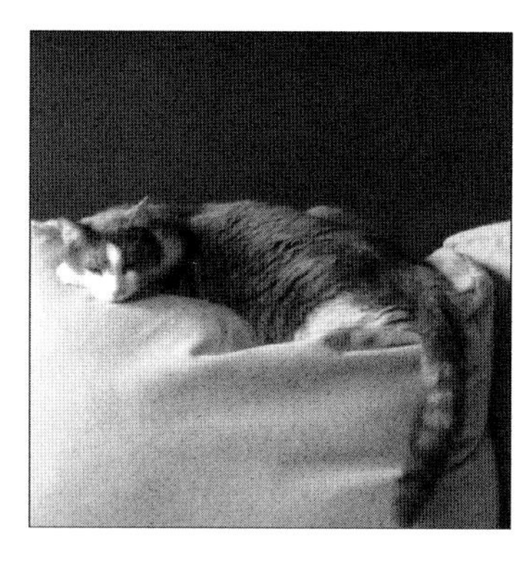

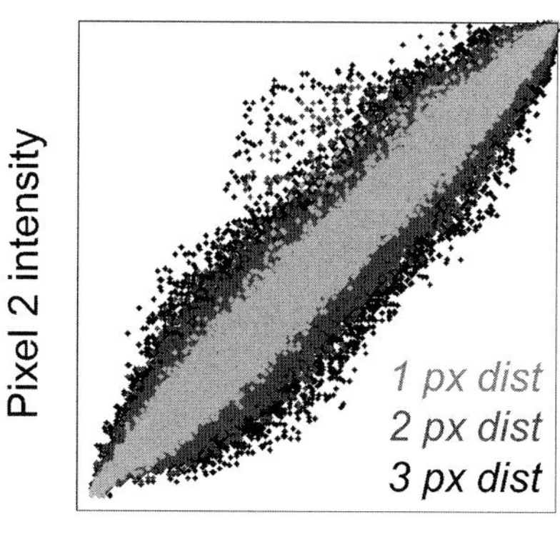


**Figure 4.** Structure in the sensory world. Sensory stimuli in the environment, such as the image of the cat, display significant statistical structure. For example, the luminance value of nearby pixels in the image are significantly correlated, an effect that exists for even nonadjacent pixels. Statistical structure in the sensory environment extends beyond simple two-point correlations, for example, to stereotyped spatial frequency characteristics. Neural systems can improve their coding efficiency by accounting for and reducing such information redundancy.

information fundamental to the field of communication.[103] Attneave applied these ideas to perception, suggesting that a guiding principle for sensory systems is the statistically efficient representation of available information.[104] Extending this idea to the neural level, Barlow proposed that the goal of early neurons in sensory processing is to remove the redundancy in the input stimuli.[105]

These approaches are motivated by the fact that signals arising from the natural environment are highly structured.[94] Such structural regularities imply informational redundancy because an observer with knowledge about part of a signal can predict other parts of the signal with greater than chance probability. Consider the processing of visual information, from which most studies of natural statistics and empirical evidence for neural efficient coding have arisen. As initially pointed out by Attneave,[104] there is a significant degree of redundancy in natural visual images because of correlation in both the spatial and temporal domains. For example, as shown in Figure 4, if the responses of a pair of pixels separated by a fixed distance are examined across all such pairs in a natural image, this activity will be highly correlated. This kind of spatial correlation is clearly evident upon even a casual examination of a natural image, and such structural redundancies underlie current image compression and transmission technologies. These correlations reflect the underlying smoothness of natural images in both space and time: luminance primarily changes gradually (with the exception of sharp transitions at edges), line segments vary as contours, and visual inputs change smoothly with time. Additional structure is evident when one considers the spatial and temporal dimensions together.[106] These statistical regularities constrain the images a visual system is likely to encounter to a tiny fraction of the set of all possible images, and visual circuits must be tuned to this probable subset in order to represent this information efficiently.

How is efficient coding evident in neural responses? At the level of single neurons, efficient coding requires that the input–output function be adjusted so that the entire response range is employed to represent the stimulus distribution.[107] For example, under the constraints of a maximum firing rate and finite precision, efficient neurons should employ all activity levels equally in response to the distribution (Fig. 5). If the input–output function sensitivity is set too low, high levels of the stimulus feature will be indistinguishable as the response function saturates; if the sensitivity is set too high, low levels of the stimulus feature cannot drive responses. One early demonstration of the precise correspondence between activity and natural stimulus statistics is Laughlin's work on contrast-sensitive neurons in the fly compound eye.[108] Much like the model depicted in Figure 5, the contrast-response function of these neurons replicates the curve that transforms the probability distribution of natural contrasts into a flat response distribution,

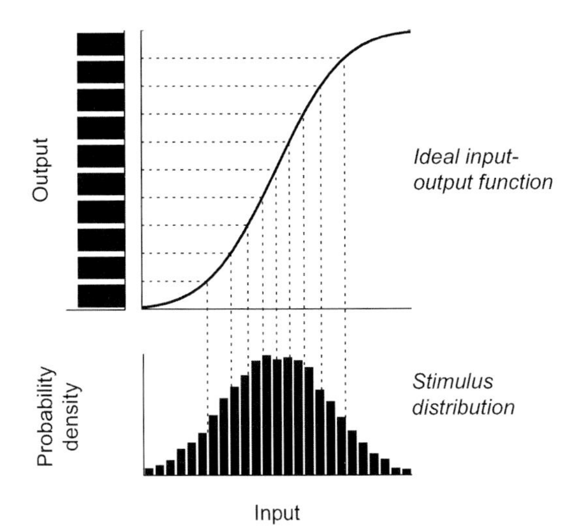


**Figure 5.** Efficient coding in sensory systems. For a given distribution of sensory characteristics in the world (bottom), an efficient neural input–output function produces an output (top) that equally uses all possible levels of neural activity. Such a function matches the greatest sensitivity of neural responses to the most probable stimuli in the environment, and has been demonstrated in neurophysiological data from sensory systems. (Adapted from Laughlin, 1981.[108])

thus matching neural activity to the environmental statistics. When the activity of multiple neurons are considered together, the efficient encoding hypothesis requires that the joint encoding of a stimulus should reflect both optimal responses in individual neurons and efficiency across the set of neurons. For example, to maximize efficiency and reduce redundancy, neural responses should be independent of one another (decorrelated), and a given stimulus should involve only a small fraction of the available neurons (sparse).

Recent work suggests that contextual effects in visual processing like surround modulation and adaptation serve to implement efficient coding representations. Theoretically, retinal and LGN center-surround structures implement a spatial decorrelation of outputs[109] (termed a *whitening* of the spectrum, in the frequency domain) and the nonlinear interactions mediated by cortical eRF interactions increase the sparseness of the output representations.[110] Efficient coding has been proposed to explain the effect of context on neural responses, like the orientation tuning curve changes driven by surround stimuli.[88,111,112] Empirically, when natural stimuli drive the surround of V1 neurons, responses are decorrelated and show a more efficient sparse representation compared to cRF stimulation alone.[113,114] To explore how contextual modulation increases the efficiency of sensory processing, Schwartz and Simoncelli explored the possible computational mechanisms linking the two. Specifically, they examined *divisive normalization*, a gain-control mechanism that characterizes contextual effects like surround suppression and contrast gain control. They found that models incorporating divisive normalization increase the independence of neural responses and allow for efficient encoding of natural visual and auditory signals.[67]

Modulation by temporal context can also serve to improve the efficiency of sensory processing. The fundamental principle of efficient coding is that a sensory system is adjusted to the specific statistics of the natural environment from which it encodes and transmits information. However, if a sensory system is hard-wired to only the global, long-term average statistics of the world, it cannot efficiently transmit information if the short-term, local statistics vary. Experimental studies such as those reviewed previously suggest that sensory systems adapt to not only the mean of stimulus distributions, but to high-order statistics such as the variance. When coding efficiency is quantified, adaptation rescales neuronal input–output functions in a manner that maximizes the transmission of information.[115,116]

## An efficient coding framework for context-dependent value encoding

How can sensory context effects and an efficient coding framework illuminate our understanding of the representation of value in neural circuits? The efficient coding hypothesis proposes that sensory systems reduce the redundancy in natural signals (by increasing the efficiency of their output responses) in order to maximize the information that can be transmitted through limited capacity channels. Unlike sensory information, value is not a product of the environment alone, but a subjective construct determined by both external information and the state of the animal. Thus, value coding systems face both external constraints given by the statistics of reward distributions in the natural world and internal constraints governed by factors such as physiological status and metabolic needs. However, research in foraging theory shows that animals in the real world behave in manners distinct and highly suited to the reward structure in their environmental niches.[53]

This suggests that the number of possible value distributions in the environment is constrained, and that the neural circuits driving behavior are tuned to this environmental structure.

We hypothesize here that value systems in the brain adopt coding strategies specific to their functional requirements, a feature that specifies the different forms of context-dependent value representation. In particular, we hypothesize that it is the statistics of the value distributions that each circuit encounters that will structure the form of the value representations. For action-value coding in decision-related systems, value representations arise simultaneously during action selection when a choice must be made between multiple options. For value systems involved in value and representation and storage, different value representations may be activated at different times depending on the environmental requirements. These response patterns are analogous to the manner in which the spatial and temporal aspects of incoming sensory signals are represented, and contextual effects in sensory processing offer a framework in which to explore the analogous context-dependent value representations.

As reviewed previously, the neural representation of action value appears to be instantiated in a relative manner, dependent on the other action values available at the time. This value representation is a spatial form of contextual modulation: analogous to extra-classical RF effects in the visual system, modulation of a given neuron is driven by stimuli that themselves do not drive the spiking activity. Although the function of the early sensory system is to efficiently transmit information about stimulus features, a decision circuit must select the best option and discriminate between the values of the possible choices. In terms of efficient coding, the simultaneous options should be represented optimally in neurons coding single actions and across the set of active neurons coding the choice set.

For a single neuron, divisive normalization acts as a means of compressive gain control even when a single option is presented alone. Because reward amounts are potentially limitless in the real world, a gain control mechanism operates to transform this wide range into the limited dynamic range of neural firing rates. Analogous to mechanisms like the one Laughlin described in the fly eye, divisive normalization may adjust the value input–output function to efficiently encode the distribution of possible values. Seen in this light, the shape of the gain control observed in area LIP may reflect an underlying value distribution that is significantly skewed, with most possible options occurring at low values and a long tail of rare, higher values. The degree to which such compression increases the efficiency of the representation is difficult to quantify without knowledge about the natural statistics of rewards. Although ecological reward distributions have been studied for some animals and niches, the distribution of natural rewards in higher order primates, particularly over evolutionary timescales, remains unknown.

However, like sensory systems, decision systems must process multiple representations at the same instant in time. For neurons encoding the value of multiple actions, contextual modulation may serve to adjust the range of neuronal firing rates across the set of neurons to the value distribution of the choice set. Consider a decision system whose goal is to distinguish the higher valued of two rewards separated by a small amount when the values are low versus when they are high. In addition to providing a compression that keeps outputs within the dynamic range of single neurons, a relative value representation dependent on the total reward available adjusts the gain across the entire set of active neurons. Significantly, the divisive normalization computation that precisely characterizes relative value representation in the LIP also underlies spatial context effects in visual processing, raising the possibility that divisive normalization represents a canonical cortical computation that drives efficient gain control. This form of instantaneous normalization across active neurons may also apply to other brain areas that simultaneously represent the value of stimuli or actions. For example, areas that serve as saliency or attentional maps, representing simultaneous information from locations spanning the environment, may also implement a divisive normalization.[69]

In the temporal domain, contextual effects in sensory processing have an analogue in the adaptation of neurons coding economic value in the OFC that reflects the proposed role of orbitofrontal neurons in the storage of value information. Unlike action-value coding neurons in the LIP or premotor cortex, OFC neurons are menu invariant and are unmodulated by the presence of other choice options. However, like neurons in the visual system that adapt to statistics of the stimulus feature distribution, OFC responses show modulation by values encountered over a longer timescale. This adaptation is sensitive to multiple distribution statistics, including the mean, range, and variance of the recent value signals. Intuitively, adaptation effects improve the efficiency of coding by adjusting the input–output function of value neurons to the appropriate local statistics of input values. For example, when presented with reward distributions with identical means but different variances, OFC neurons adapt their responses so that their reward sensitivity slopes align with the probable environmental rewards. Using mutual information theory to quantify the ability of neurons to discriminate rewards, Kobayashi *et al.* showed that adaptive neurons in OFC preserve the amount of encoded information regardless of the input statistics.[75] Thus, adaptive processes in value storage areas may produce a more efficient neural representation of value for use in downstream decision processes.

## Caveats

There remains much that is unknown about valuation systems in the brain and their relationship to the environment. Unlike natural sensory signals, the distribution of values in the natural environment is not nearly as easily defined or measured, particularly if one wishes to examine the statistics of values over an evolutionary timescale. It is comparatively easy to extract the statistics of the sensory environment, but the statistics of value will always be subject to the interaction between an organism and its environment. Experimenters are beginning to address this issue in the laboratory by constraining the statistics of local values, an approach particularly suited to the study of adaptive processes.

The parallels between contextual effects in value coding and sensory processing reviewed previously may reflect a shared functional architecture (like divisive normalization), a shared design principle (dependence on natural statistics), or both. Although the principle of efficient coding in sensory processing provides an attractive framework to examine contextual effects in value representation, it is important to note that the sensory system and valuation networks have different functional goals. Information theory approaches representation as purely a problem of transmission, concerned with maximizing the amount of information in the signal while reducing redundancy. This approach has reliably characterized many of the early stages of sensory processing, in which the primary goal is to transmit information about the environment to the rest of the brain, but higher order brain areas are likely to have functional goals beyond strict transmission. Ultimately, the selection pressure for all neural systems, including sensory ones, is not maximizing the efficient representation of information *per se* but maximizing the survival of the animal. Given the high metabolic cost of neural systems, however, evolution should favor an optimal use of resources, regardless of whether this is implemented as efficiency in representation, as proposed by the efficient-coding hypothesis, or as another constraint. It is likely that efficient coding will not be the only principle through which to understand the design of value systems, but it provides an attractive starting point to examine the nature of value processing.

## Conflicts of interest

The authors declare no conflicts of interest.

## References


1. Von Neumann, J. & O. Morgenstern. 1944. *Theory of Games and Economic Behavior*. Princeton University Press. Princeton, NJ.
2. Samuelson, P.A. 1947. *Foundations of Economic Analysis*. Harvard University Press. Cambridge.
3. Kable, J.W. & P.W. Glimcher. 2007. The neural correlates of subjective value during intertemporal choice. *Nat. Neurosci.* **10:** 1625–1633.
4. Kable, J.W. & P.W. Glimcher. 2009. The neurobiology of decision: consensus and controversy. *Neuron* **63:** 733–745.
5. Rangel, A. & T. Hare. 2010. Neural computations associated with goal-directed choice. *Curr. Opin. Neurobiol.* **20:** 262–270.
6. Platt, M.L. & P.W. Glimcher. 1999. Neural correlates of decision variables in parietal cortex. *Nature* **400:** 233–238.
7. Dorris, M.C. & P.W. Glimcher. 2004. Activity in posterior parietal cortex is correlated with the relative subjective desirability of action. *Neuron* **44:** 365–378.
8. Sugrue, L.P., G.S. Corrado & W.T. Newsome. 2004. Matching behavior and the representation of value in the parietal cortex. *Science* **304:** 1782–1787.
9. Seo, H., D.J. Barraclough & D. Lee. 2009. Lateral intraparietal cortex and reinforcement learning during a mixed-strategy game. *J. Neurosci.* **29:** 7278–7289.
10. Louie, K. & P.W. Glimcher. 2010. Separating value from choice: delay discounting activity in the lateral intraparietal area. *J. Neurosci.* **30:** 5498–5507.
11. Ding, L. & O. Hikosaka. 2006. Comparison of reward modulation in the frontal eye field and caudate of the macaque. *J. Neurosci.* **26:** 6695–6703.

12. So, N.Y. & V. Stuphorn. 2010. Supplementary eye field encodes option and action value for saccades with variable reward. *J. Neurophysiol.* **104:** 2634–2653.
13. Dorris, M.C. & D.P. Munoz. 1998. Saccadic probability influences motor preparation signals and time to saccadic initiation. *J. Neurosci.* **18:** 7015–7026.
14. Ikeda, T. & O. Hikosaka. 2003. Reward-dependent gain and bias of visual responses in primate superior colliculus. *Neuron* **39:** 693–700.
15. Hollerman, J.R., L. Tremblay & W. Schultz. 1998. Influence of reward expectation on behavior-related neuronal activity in primate striatum. *J. Neurophysiol.* **80:** 947–963.
16. Kawagoe, R., Y. Takikawa & O. Hikosaka. 1998. Expectation of reward modulates cognitive signals in the basal ganglia. *Nat. Neurosci.* **1:** 411–416.
17. Cromwell, H.C. & W. Schultz. 2003. Effects of expectations for different reward magnitudes on neuronal activity in primate striatum. *J. Neurophysiol.* **89:** 2823–2838.
18. Samejima, K. *et al.* 2005. Representation of action-specific reward values in the striatum. *Science* **310:** 1337–1340.
19. Lau, B. & P.W. Glimcher. 2008. Value representations in the primate striatum during matching behavior. *Neuron* **58:** 451–463.
20. Musallam, S. *et al.* 2004. Cognitive control signals for neural prosthetics. *Science* **305:** 258–262.
21. Hori, Y., T. Minamimoto & M. Kimura. 2009. Neuronal encoding of reward value and direction of actions in the primate putamen. *J. Neurophysiol.* **102:** 3530–3543.
22. Padoa-Schioppa, C. & J.A. Assad. 2006. Neurons in the orbitofrontal cortex encode economic value. *Nature* **441:** 223–226.
23. Padoa-Schioppa, C. 2011. Neurobiology of economic choice: a good-based model. *Annu. Rev. Neurosci.* **34:** 333–359.
24. Thorpe, S.J., E.T. Rolls & S. Maddison. 1983. The orbitofrontal cortex: neuronal activity in the behaving monkey. *Exp. Brain Res.* **49:** 93–115.
25. Tremblay, L. & W. Schultz. 1999. Relative reward preference in primate orbitofrontal cortex. *Nature* **398:** 704–708.
26. Wallis, J.D. & E.K. Miller. 2003. Neuronal activity in primate dorsolateral and orbital prefrontal cortex during performance of a reward preference task. *Eur. J. Neurosci.* **18:** 2069–2081.
27. Roesch, M.R. & C.R. Olson. 2004. Neuronal activity related to reward value and motivation in primate frontal cortex. *Science* **304:** 307–310.
28. Ongur, D. & J.L. Price. 2000. The organization of networks within the orbital and medial prefrontal cortex of rats, monkeys and humans. *Cereb. Cortex* **10:** 206–219.
29. Paton, J.J. *et al.* 2006. The primate amygdala represents the positive and negative value of visual stimuli during learning. *Nature* **439:** 865–870.
30. Belova, M.A., J.J. Paton & C.D. Salzman. 2008. Moment-to-moment tracking of state value in the amygdala. *J. Neurosci.* **28:** 10023–10030.
31. Wise, R.A. 1996. Addictive drugs and brain stimulation reward. *Annu. Rev. Neurosci.* **19:** 319–340.
32. Wise, R.A. 2004. Dopamine, learning and motivation. *Nat. Rev. Neurosci.* **5:** 483–494.
33. Romo, R. & W. Schultz. 1990. Dopamine neurons of the monkey midbrain: contingencies of responses to active touch during self-initiated arm movements. *J. Neurophysiol.* **63:** 592–606.
34. Ljungberg, T., P. Apicella & W. Schultz. 1992. Responses of monkey dopamine neurons during learning of behavioral reactions. *J. Neurophysiol.* **67:** 145–163.
35. Mirenowicz, J. & W. Schultz. 1994. Importance of unpredictability for reward responses in primate dopamine neurons. *J. Neurophysiol.* **72:** 1024–1027.
36. Hollerman, J.R. & W. Schultz. 1998. Dopamine neurons report an error in the temporal prediction of reward during learning. *Nat. Neurosci.* **1:** 304–309.
37. Montague, P.R. *et al.* 1995. Bee foraging in uncertain environments using predictive hebbian learning. *Nature* **377:** 725–728.
38. Montague, P.R., P. Dayan & T.J. Sejnowski. 1996. A framework for mesencephalic dopamine systems based on predictive Hebbian learning. *J. Neurosci.* **16:** 1936–1947.
39. Glimcher, P.W. 2011. Understanding dopamine and reinforcement learning: the dopamine reward prediction error hypothesis. *Proc. Natl. Acad. Sci. USA* **108**(Suppl 3): 15647–15654.
40. Kalman, R.E. 1960. A new approach to linear filtering and prediction problems. *J. Basic Eng. Trans. ASME* **82:** 35–45.
41. Dickinson, A. 1980. *Contemporary Animal Learning Theory*. Cambridge University Press. Cambridge [Eng.]; New York.
42. Sutton, R.S. & A.G. Barto. 1981. Toward a modern theory of adaptive networks: expectation and prediction. *Psychol. Rev.* **88:** 135–170.
43. Sutton, R.S. & A.G. Barto. 1998. *Reinforcement Learning: An Introduction*. MIT Press. Cambridge, MA.
44. Schultz, W., P. Dayan & P.R. Montague. 1997. A neural substrate of prediction and reward. *Science* **275:** 1593–1599.
45. Waelti, P., A. Dickinson & W. Schultz. 2001. Dopamine responses comply with basic assumptions of formal learning theory. *Nature* **412:** 43–48.
46. Bayer, H.M. & P.W. Glimcher. 2005. Midbrain dopamine neurons encode a quantitative reward prediction error signal. *Neuron* **47:** 129–141.
47. Glimcher, P.W. 2011. *Foundations of Neuroeconomic Analysis*. Oxford University Press. New York.
48. Maunsell, J.H. 2004. Neuronal representations of cognitive state: reward or attention? *Trends Cogn Sci.* **8:** 261–265.
49. Gottlieb, J.P., M. Kusunoki & M.E. Goldberg. 1998. The representation of visual salience in monkey parietal cortex. *Nature* **391:** 481–484.
50. Bisley, J.W. & M.E. Goldberg. 2003. Neuronal activity in the lateral intraparietal area and spatial attention. *Science* **299:** 81–86.
51. Bisley, J.W. & M.E. Goldberg. 2010. Attention, intention, and priority in the parietal lobe. *Annu. Rev. Neurosci.* **33:** 1–21.
52. Luce, R.D. 1959. *Individual Choice Behavior: A Theoretical Analysis*. Wiley. New York.

53. Stephens, D.W. & J.R. Krebs. 1986. *Foraging Theory*. Princeton University Press. Princeton, NJ.
54. Herrnstein, R.J. 1961. Relative and absolute strength of response as a function of frequency of reinforcement. *J. Exp. Anal. Behav.* **4:** 267–272.
55. Shadlen, M.N. & W.T. Newsome. 2001. Neural basis of a perceptual decision in the parietal cortex (area LIP) of the rhesus monkey. *J. Neurophysiol.* **86:** 1916–1936.
56. Toth, L.J. & J.A. Assad. 2002. Dynamic coding of behaviourally relevant stimuli in parietal cortex. *Nature* **415:** 165–168.
57. Leon, M.I. & M.N. Shadlen. 2003. Representation of time by neurons in the posterior parietal cortex of the macaque. *Neuron* **38:** 317–327.
58. Yang, T. & M.N. Shadlen. 2007. Probabilistic reasoning by neurons. *Nature* **447:** 1075–1080.
59. Rorie, A.E. *et al.* 2010. Integration of sensory and reward information during perceptual decision-making in lateral intraparietal cortex (LIP) of the macaque monkey. *PLoS One* **5:** e9308.
60. Roitman, J.D. & M.N. Shadlen. 2002. Response of neurons in the lateral intraparietal area during a combined visual discrimination reaction time task. *J. Neurosci.* **22:** 9475–9489.
61. Heeger, D.J. 1992. Normalization of cell responses in cat striate cortex. *Vis. Neurosci.* **9:** 181–197.
62. Carandini, M., D.J. Heeger & J.A. Movshon. 1997. Linearity and normalization in simple cells of the macaque primary visual cortex. *J. Neurosci.* **17:** 8621–8644.
63. Britten, K.H. & H.W. Heuer. 1999. Spatial summation in the receptive fields of MT neurons. *J. Neurosci.* **19:** 5074–5084.
64. Cavanaugh, J.R., W. Bair & J.A. Movshon. 2002. Nature and interaction of signals from the receptive field center and surround in macaque V1 neurons. *J. Neurophysiol.* **88:** 2530–2546.
65. Heuer, H.W. & K.H. Britten. 2002. Contrast dependence of response normalization in area MT of the rhesus macaque. *J. Neurophysiol.* **88:** 3398–3408.
66. Zoccolan, D., D.D. Cox & J.J. DiCarlo. 2005. Multiple object response normalization in monkey inferotemporal cortex. *J. Neurosci.* **25:** 8150–8164.
67. Schwartz, O. & E.P. Simoncelli. 2001. Natural signal statistics and sensory gain control. *Nat. Neurosci.* **4:** 819–825.
68. Valerio, R. & R. Navarro. 2003. Optimal coding through divisive normalization models of V1 neurons. *Network* **14:** 579–593.
69. Reynolds, J.H. & D.J. Heeger. 2009. The normalization model of attention. *Neuron* **61:** 168–185.
70. Carandini, M. & D.J. Heeger. 2012. Normalization as a canonical neural computation. *Nat. Rev. Neurosci.* **13:** 51–62.
71. Louie, K., L.E. Grattan & P.W. Glimcher. 2011. Reward value-based gain control: divisive normalization in parietal cortex. *J. Neurosci.* **31:** 10627–10639.
72. Pastor-Bernier, A. & P. Cisek. 2011. Neural correlates of biased competition in premotor cortex. *J. Neurosci.* **31:** 7083–7088.
73. Padoa-Schioppa, C. & J.A. Assad. 2008. The representation of economic value in the orbitofrontal cortex is invariant for changes of menu. *Nat. Neurosci.* **11:** 95–102.
74. Padoa-Schioppa, C. 2009. Range-adapting representation of economic value in the orbitofrontal cortex. *J. Neurosci.* **29:** 14004–14014.
75. Kobayashi, S., O. Pinto de Carvalho & W. Schultz. 2010. Adaptation of reward sensitivity in orbitofrontal neurons. *J. Neurosci.* **30:** 534–544.
76. Tobler, P.N., C.D. Fiorillo & W. Schultz. 2005. Adaptive coding of reward value by dopamine neurons. *Science* **307:** 1642–1645.
77. Samuelson, P.A. 1947. *Foundations of Economic Analysis*. Harvard University Press. Cambridge, MA.
78. Tversky, A. 1972. Elimination by aspects: a theory of choice. *Psychol. Rev.* **79:** 281-&.
79. Huber, J., J.W. Payne & C. Puto. 1982. Adding asymmetrically dominated alternatives: violations of regularity and the similarity hypothesis. *J. Consum. Res.* **9:** 90–98.
80. Shafir, S., T.A. Waite & B.H. Smith. 2002. Context-dependent violations of rational choice in honeybees (*Apis mellifera*) and gray jays (*Perisoreus canadensis*). *Behav. Ecol. Sociobiol.* **51:** 180–187.
81. Bateson, M., S.D. Healy & T.A. Hurly. 2003. Context-dependent foraging decisions in rufous hummingbirds. *P. Roy. Soc. Lond. B Bio.* **270:** 1271–1276.
82. Tversky, A. & E. Shafir. 1992. Choice under conflict: the dynamics of deferred decision. *Psychol. Sci.* **3:** 358–361.
83. Iyengar, S.S. & M.R. Lepper. 2000. When choice is demotivating: can one desire too much of a good thing? *J. Pers. Soc. Psychol.* **79:** 995–1006.
84. DeShazo, J.R. & G. Fermo. 2002. Designing choice sets for stated preference methods: the effects of complexity on choice consistency. *J. Environ. Econ. Manag.* **44:** 123–143.
85. Tolhurst, D.J., J.A. Movshon & A.F. Dean. 1983. The statistical reliability of signals in single neurons in cat and monkey visual cortex. *Vision Res.* **23:** 775–785.
86. Softky, W.R. & C. Koch. 1993. The highly irregular firing of cortical cells is inconsistent with temporal integration of random EPSPs. *J. Neurosci.* **13:** 334–350.
87. Shadlen, M.N. & W.T. Newsome. 1998. The variable discharge of cortical neurons: implications for connectivity, computation, and information coding. *J. Neurosci.* **18:** 3870–3896.
88. Schwartz, O., A. Hsu & P. Dayan. 2007. Space and time in visual context. *Nat. Rev. Neurosci.* **8:** 522–535.
89. Albright, T.D. & G.R. Stoner. 2002. Contextual influences on visual processing. *Annu. Rev. Neurosci.* **25:** 339–379.
90. DeAngelis, G.C., R.D. Freeman & I. Ohzawa. 1994. Length and width tuning of neurons in the cat's primary visual cortex. *J. Neurophysiol.* **71:** 347–374.
91. Gilbert, C.D. & T.N. Wiesel. 1985. Intrinsic connectivity and receptive field properties in visual cortex. *Vision Res.* **25:** 365–374.
92. Bair, W., J.R. Cavanaugh & J.A. Movshon. 2003. Time course and time-distance relationships for surround suppression in macaque V1 neurons. *J. Neurosci.* **23:** 7690–7701.
93. Ozeki, H. *et al.* 2004. Relationship between excitation and inhibition underlying size tuning and contextual response

modulation in the cat primary visual cortex. *J. Neurosci.* **24:** 1428–1438.
94. Simoncelli, E.P. & B.A. Olshausen. 2001. Natural image statistics and neural representation. *Annu. Rev. Neurosci.* **24:** 1193–1216.
95. Shapley, R.M. & C. Enroth-Cugell. 1984. Visual adaptation and retinal gain control. *Prog. Ret. Res.* **3:** 263–346.
96. Kohn, A. 2007. Visual adaptation: physiology, mechanisms, and functional benefits. *J. Neurophysiol.* **97:** 3155–3164.
97. Ohzawa, I., G. Sclar & R.D. Freeman. 1982. Contrast gain control in the cat visual cortex. *Nature* **298:** 266–268.
98. Smirnakis, S.M. *et al.* 1997. Adaptation of retinal processing to image contrast and spatial scale. *Nature* **386:** 69–73.
99. Bonin, V., V. Mante & M. Carandini. 2006. The statistical computation underlying contrast gain control. *J. Neurosci.* **26:** 6346–6353.
100. Sharpee, T.O. *et al.* 2006. Adaptive filtering enhances information transmission in visual cortex. *Nature* **439:** 936–942.
101. Maravall, M. *et al.* 2007. Shifts in coding properties and maintenance of information transmission during adaptation in barrel cortex. *PLoS Biol.* **5:** e19.
102. Nagel, K.I. & A.J. Doupe. 2006. Temporal processing and adaptation in the songbird auditory forebrain. *Neuron* **51:** 845–859.
103. Shannon, C.E. & W. Weaver. 1949. *The Mathematical Theory of Communication.* University of Illinois Press. Urbana, IL.
104. Attneave, F. 1954. Some informational aspects of visual perception. *Psychol. Rev.* **61:** 183–193.
105. Barlow, H.B. 1961. Possible principles underlying the transformation of sensory messages. In *Sensory Communication.* W.A. Rosenblith, Ed. MIT Press. Cambridge, MA.
106. van Hateren, J.H. & D.L. Ruderman. 1998. Independent component analysis of natural image sequences yields spatio-temporal filters similar to simple cells in primary visual cortex. *Proc. Biol. Sci.* **265:** 2315–2320.
107. Simoncelli, E.P. 2003. Vision and the statistics of the visual environment. *Curr. Opin. Neurobiol.* **13:** 144–149.
108. Laughlin, S. 1981. A simple coding procedure enhances a neuron's information capacity. *Z. Naturforsch. C.* **36:** 910–912.
109. Atick, J.J. 1992. Could information theory provide an ecological theory of sensory processing? *Network* **3:** 213–251.
110. Olshausen, B.A. & D.J. Field. 1997. Sparse coding with an overcomplete basis set: a strategy employed by V1? *Vision Res.* **37:** 3311–3325.
111. Muller, J.R. *et al.* 2003. Local signals from beyond the receptive fields of striate cortical neurons. *J. Neurophysiol.* **90:** 822–831.
112. Felsen, G., J. Touryan & Y. Dan. 2005. Contextual modulation of orientation tuning contributes to efficient processing of natural stimuli. *Network* **16:** 139–149.
113. Vinje, W.E. & J.L. Gallant. 2000. Sparse coding and decorrelation in primary visual cortex during natural vision. *Science* **287:** 1273–1276.
114. Vinje, W.E. & J.L. Gallant. 2002. Natural stimulation of the nonclassical receptive field increases information transmission efficiency in V1. *J. Neurosci.* **22:** 2904–2915.
115. Brenner, N., W. Bialek & R. de Ruyter van Steveninck. 2000. Adaptive rescaling maximizes information transmission. *Neuron* **26:** 695–702.
116. Fairhall, A.L. *et al.* 2001. Efficiency and ambiguity in an adaptive neural code. *Nature* **412:** 787–792.

Ann. N.Y. Acad. Sci. ISSN 0077-8923

ANNALS OF THE NEW YORK ACADEMY OF SCIENCES
Issue: *The Year in Cognitive Neuroscience*

# The emotion paradox in the aging brain

Mara Mather
The USC Davis School of Gerontology, University of Southern California, Los Angeles, California

Address for correspondence: Mara Mather, Ph.D., The USC Davis School of Gerontology, University of Southern California, 3715 McClintock Ave., Los Angeles, CA 90089. mara.mather@usc.edu

This paper reviews age differences in emotion processing and how they may relate to age-related changes in the brain. Compared with younger adults, older adults react less to negative situations, ignore irrelevant negative stimuli better, and remember relatively more positive than negative information. Older adults' ability to insulate their thoughts and emotional reactions from negative situations is likely due to a number of factors, such as being less influenced by interoceptive cues, selecting different emotion regulation strategies, having less age-related decline in prefrontal regions associated with emotional control than in other prefrontal regions, and engaging in emotion regulation strategies as a default mode in their everyday lives. Healthy older adults' avoidance of processing negative stimuli may contribute to their well-maintained emotional well-being. However, when cardiovascular disease leads to additional prefrontal white matter damage, older adults have fewer cognitive control mechanisms available to regulate their emotions, making them more vulnerable to depression. In general, although age-related changes in the brain help shape emotional experience, shifts in preferred strategies and goal priorities are also important influences.

**Keywords:** emotion regulation; aging; fMRI; ventromedial prefrontal cortex; amygdala

## Introduction

Emotions prepare the body for action, guide decisions, and highlight what should be attended to and remembered. However, it can be difficult to turn off or ignore emotions. When negative emotions last for a long time and are maintained even when no longer appropriate, disorders such as chronic depression or anxiety ensue. As emotions are central to everyday function and well-being, it is important to understand how aging affects the experience and regulation of emotion. Recent research reveals intriguing findings about emotion and aging and raises questions about which aspects of these age-related changes in emotional processing can be accounted for by age-related changes in the brain, and which are more influenced by other factors, such as changes in other body systems, increased experience, or shifts in goals. In this paper, I review various types of age differences in emotion experience and regulation, and discuss relevant changes in the brain.

## The paradox of emotional well-being in aging

In old age, physical health and strength typically decline. The death of a spouse or close friends is more likely than earlier in life. Cognitive agility decreases and it is harder to remember information. Social networks shrink. This picture sounds depressing. Yet, on average, adults' emotional well-being is somehow maintained or even improved as they age.[1] For instance, older adults move out of highly negative emotional states faster than younger adults do and maintain the absence of negative affect more consistently.[2,3] Middle-aged and older adults are less physically and emotionally reactive to interpersonal stressors than younger adults.[4] When older adults experience interpersonal tensions, they engage less in destructive conflict strategies, such as yelling, arguing, or name calling,[5] and generally find tense interpersonal situations less stressful than younger adults do.[6]

doi: 10.1111/j.1749-6632.2012.06471.x

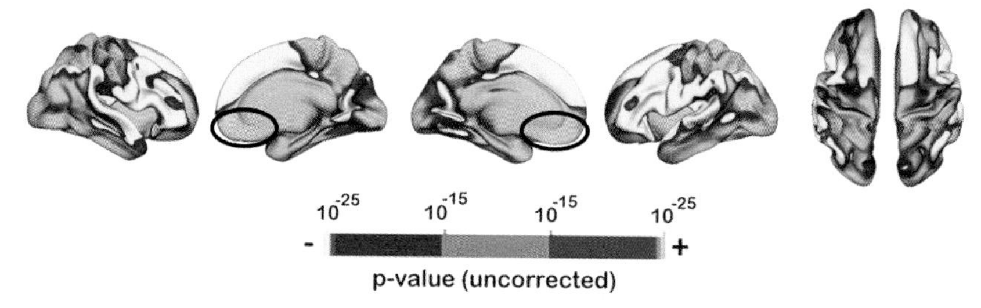


**Figure 1.** Fjell *et al.*[10] assessed cortical thickness across six different samples with a total of 883 participants. Negative correlations between age and cortical thickness were seen in the superior, middle, and inferior frontal gyri, but not in the anterior cingulate cortices or in ventromedial prefrontal cortex. The frontal regions that showed little relationship with age are circled in the image. Figure is adapted from Fjell *et al.*[10]

What might explain older adults' pattern of lower reactivity to negative situations and more ephemeral negative emotional experience? There are two obvious potential answers to this puzzle. One is that older adults are better than younger adults at regulating their emotions and so can more easily defuse negative feelings and situations. The other is that older adults do not experience negative feelings and situations as intensely as younger adults do and so have less need to defuse them. In this paper, I will review behavioral and brain data to critically examine both of these putative explanations for the paradox of emotion in aging.

## Aging affects dorsal and lateral PFC more than ventromedial PFC

To answer the question of whether older adults differ from younger adults in their abilities to regulate emotions, it is important to consider whether there are age-related declines in the brain regions associated with emotion regulation abilities. Research indicates that effective emotion regulation relies on the prefrontal cortex (PFC). Both the ventromedial PFC and the adjacent anterior cingulate cortex (ACC) play important roles in controlling one's own emotion. For instance, close family members rated ventromedial PFC lesion patients as more irritable, emotionally labile, apathetic, and emotionally impoverished than did family members of nonventromedial PFC lesion patients.[7] Across both the ACC and the PFC, ventral regions are more critical for emotional processing, whereas dorsal regions are more critical for cognitive executive functions.[8]

Although the PFC is especially vulnerable to decline in normal aging, this vulnerability varies by subregion. In particular, the ventromedial PFC develops earlier in childhood than other regions[9] and maintains its cortical thickness throughout the adult life span[10] (Fig. 1). The ACC also maintains its cortical thickness in normal aging, whereas lateral and superior regions of PFC thin dramatically. Findings that cortical thickness declines more in aging in dorsal than ventral ACC and PFC (Fig. 1) suggest that some of the neural circuitry critical for emotion regulation is well preserved in aging. However, as reviewed in the following sections, emotion regulation processes do not remain constant, but instead change in a variety of ways with age.

## Older adults use different emotion regulation strategies than younger adults do

Which emotion regulation strategies people favor shifts with age; older adults are more likely to select strategies that involve ignoring things that might elicit negative emotion and are less likely to directly engage with emotionally negative stimuli, problems, or situations. For instance, a study that used random-digit dialing to ask over a thousand younger, middle-aged, and older Americans about their emotion regulation strategies[11] found that self-reported suppression increased with age (although the age increase was only significant for women), whereas self-reported rumination, reappraisal, and active coping all decreased with age for both genders (self-reported suppression also increased across age cohorts in a Spanish sample[12]).[a] When attempting to deal with everyday problems,

[a]In contrast with these findings of greater suppression across older cohorts, a study in Hong Kong found no age

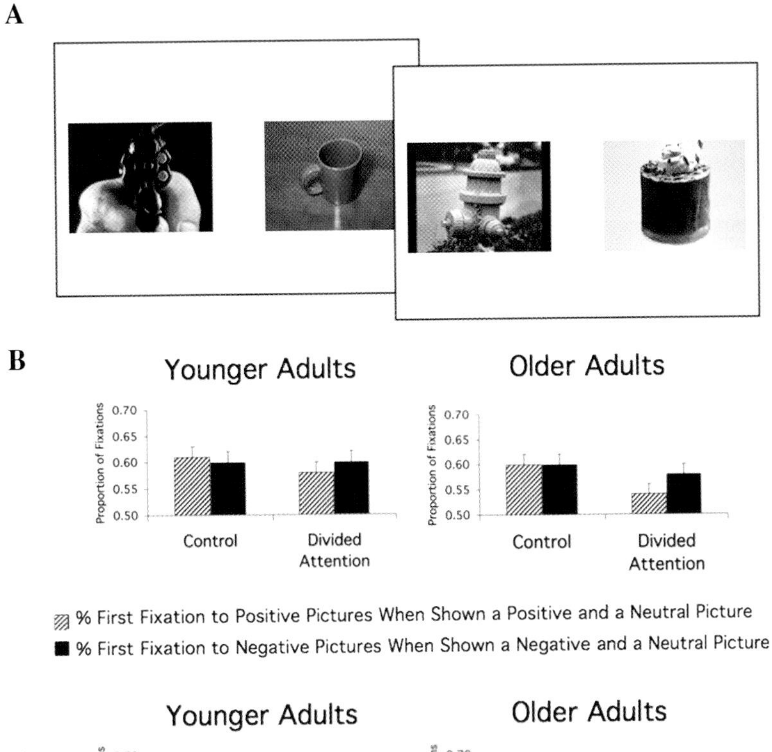


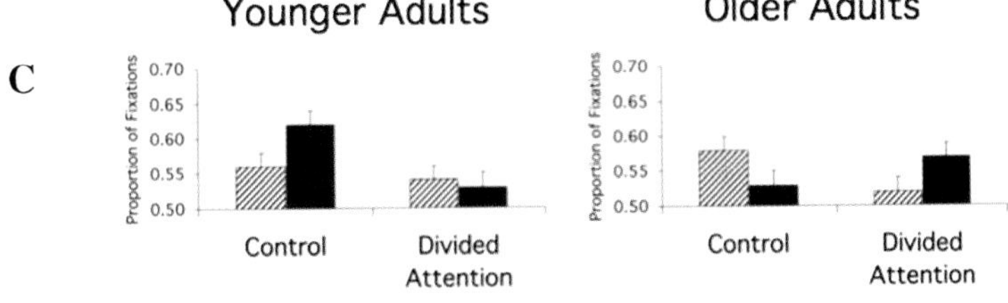


**Figure 2.** In a study by Knight *et al.*,[17] younger and older adults' eye movements were measured while they looked at pairs of pictures for six seconds each. (A) An example negative–neutral pair and an example neutral–positive pair are shown. Divided attention participants were distracted by a concurrent listening task while they looked at the pictures whereas control participants just looked at the pictures. (B) Participants' first fixation was more likely to be on an emotional picture than on a neutral picture, regardless of age. However, age differences emerged in the remaining time the pictures were shown, with older adults showing a positivity effect in the control condition but a negativity effect in the divided attention condition. Figure is from Mather.[19]

older adults are more likely to endorse strategies that involve withdrawing from the situation and attending selectively to things other than the situation itself.[14] When shown negative and neutral stimuli together, older adults tend to look away from the negative images[15–18] (Fig. 2). Furthermore, older adults show this negativity avoidance more when induced into a negative mood whereas younger adults showed mood-congruent looking.[20] These findings suggest that older adults are more likely than younger adults to ignore stimuli that might increase negative moods, and that their selective attention serves as an emotion regulation strategy.

The emotion regulation strategies people select may shift with age because of changes in which ones are easiest to implement.[21] Different aspects of

differences in suppression and greater reappraisal among middle-aged than younger adults.[13] But in this study the younger adults were university students and the middle-aged adults were recruited from local insurance companies, which may have led to group differences unrelated to age.

emotion regulation tap different PFC circuits.[22] For instance, effortful regulation often involves lateral PFC regions, whereas automatic regulation is more likely to involve ventral ACC or medial PFC.[23,24] An example of an effortful regulation strategy is when people reinterpret situations or contexts of stimuli to change their emotional meaning. During this type of reappraisal, people reliably activate dorsal PFC.[25] Thus, older adults may avoid using reappraisal to regulate emotions because it relies on regions of PFC particularly affected by aging.

## Older adults are more effective than younger adults at avoiding negative distraction

Older adults' reliance on emotion regulation strategies that involve self-directed selective attention and suppression is surprising when considering that, compared with younger adults, older adults tend to be more susceptible to distraction.[26,27] But recent research reveals an interesting pattern—although they are more easily distracted than younger adults by neutral stimuli, older adults are less easily distracted by emotional (especially negative) stimuli. For instance, a study with over 400 participants recruited via random digit dialing found that those in their 20s, 30s, and 40s were slower to name the ink color of emotional words than of neutral words, but this effect diminished with age and eventually reversed by the 70s.[28,b] In another emotional Stroop study, younger and older adults showed similar impairment in naming the color of negative words, but younger adults were also slower in naming the color of neutral words following negative words, whereas older adults did not show this sustained effect of negative information.[30] When participants were asked to indicate whether a face was fearful or happy, while ignoring the words "FEAR" or "HAPPY" superimposed on the face,[31] older adults did not perform significantly worse than younger adults on measures of conflict adaptation and interference reduction (these measures were assessed by comparing trials that had been preceded by congruent versus incongruent trials—people tend to adapt to the congruency on one trial, affecting their performance on the next trial). But when asked to indicate whether a face was male or female while ignoring the words "MALE" or "FEMALE," older adults exhibited poorer performance than the younger adults.

Although the studies described above show that older adults experience less interference from emotional stimuli than from neutral stimuli, they either did not distinguish positive from negative stimuli in their reported results or only compared negative stimuli to neutral stimuli. However, a couple of studies suggest that older adults' superior distraction avoidance is specific to negative stimuli, not positive stimuli. For instance, when asked to identify a target digit printed on top of a distractor face, younger adults were most distracted by angry faces, whereas older adults were most distracted by happy faces.[32] In another study, when asked to indicate whether two digits placed on either side of a word were both odd or even, younger adults were slower when distractor words were negative than when they were neutral or positive.[33] In contrast, older adults showed no significant reaction time differences based on the valence of the distractors. In a surprise recognition memory test 10 minutes later, younger adults showed the best recognition of the negative distractors whereas older adults showed the best recognition of the positive distractors (Fig. 3), suggesting age differences in which distractor items were attended enough to be encoded successfully.

The increasing ability with age to avoid negative distraction seems to extend to everyday life as well. In a sample of older adults between the ages of 70 and 97 who reported on their daily stressors, such as arguments, health-related problems, and how much

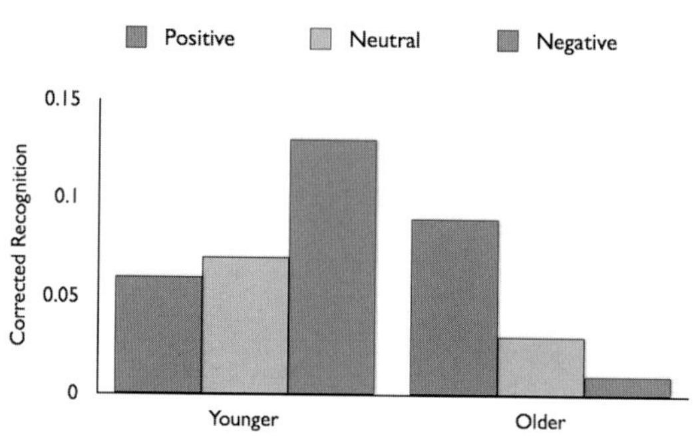


**Figure 3.** Corrected recognition (hits–false alarms) for positive, neutral, and negative words that were presented previously as irrelevant distracting items between two task-relevant digits. Figure is adapted from data presented in Thomas and Hasher.[33]

[b]Note that another study testing 36 younger and 36 older adults found conflicting results.[28] The reason for the discrepancy is not clear.

they experienced cognitive interference (e.g., "Did you think about something you didn't mean to?"), the incidence of stressful events in the past 24 hours predicted cognitive interference less with increasing age.[34]

What are the brain regions involved in ignoring negative distraction? There is evidence that, in younger adults, compared with avoiding distraction from neutral stimuli, avoiding distraction from emotional stimuli is more associated with activity in lateral inferior frontal gyrus,[35–37] and that suppressing emotional reactions to stimuli yields activity in inferior frontal gyrus as well.[22] Activity in the inferior frontal gyrus during presentation of emotionally distracting images also predicts better working memory performance for the target nonemotional memoranda,[35] suggesting that activity in this region helps diminish emotional interference. Consistent with the possibility that this region plays a critical role in avoiding emotional interference, patients with PFC brain lesions involving the inferior frontal gyrus (including Brodmann areas 44, 45, and 47) were more distracted by task-irrelevant emotional images than were patients with ventromedial PFC lesions.[38] In addition, resolving emotional interference engages regions of the ACC that are more ventral than the dorsal ACC regions engaged while resolving nonemotional interference.[39,40]

However, the precise brain regions involved in ignoring negative distractions may change with age. For example, one fMRI study suggests that the inferior frontal gyrus is differentially involved in emotional interference for younger and older adults.[41] In this study, younger and older adults were scanned as they completed a task in which they either judged the valence (negative vs. positive) or the substance (metal vs. fruit) of a central word. Some trials of each type involved congruent and others involved incongruent flanking words. The behavioral findings were consistent with those described earlier, with older adults showing more interference from incongruent words in the substance judgment task than younger adults, but less interference than younger adults in the valence judgment task. Although it did not achieve a significant interaction, there was a parallel pattern in a region of left inferior frontal gyrus; older adults showed significantly greater activity during incongruent trials than control trials in the substance but not the valence task and vice versa for younger adults. Unfortunately, from these data we cannot be sure whether more activity in this region indicates less efficient processing or more effective involvement of this region in avoiding emotional distraction. In this study, although older adults were overall better at avoiding emotional distraction than younger adults, both younger and older adults showed more distractibility from incongruent distractors that were negative than from those that were positive. No analyses by valence were reported for the fMRI data.

Whereas older adults did not have difficulty avoiding distraction from positive incongruent stimuli in the study just described, another fMRI study[42] was more consistent with the behavioral studies described earlier in which older adults were worse at avoiding positive distraction. In this study, older adults showed greater distractibility from irrelevant positive faces than from irrelevant neutral faces when attention to the central face location was high, whereas younger adults did not show differential distractibility for positive versus neutral faces. This age by distractor type interaction in reaction type was paralleled by an interaction in the ACC, such that older adults showed greater ACC activity during happy distractors shown under high attention than during neutral distractors, but younger adults did not. Furthermore, greater ACC signal during processing happy faces than neutral faces was associated with a measure of emotional stability in the older adults. No significant ACC engagement was seen in conditions with fearful or sad distractor faces. The greater ACC activation among older adults on trials with distracting positive faces may reflect both greater subjective salience of the positive distractors and the increased regulation demands imposed by this affective conflict.

In summary, older adults are often better than younger adults at ignoring emotionally negative distraction. These findings are striking given the backdrop of age-related impairments in avoiding distraction by neutral stimuli.[26,27] As suggested earlier, such findings raise the possibility that brain regions necessary for resolving interference from emotional stimuli may decline less with age than other brain regions. However, one additional critical piece of the picture is that older adults tend to be *more* distracted than younger adults by positive stimuli (although not always[41]). Thus, if their enhanced ability to ignore negative stimuli is due to relative preservation of certain PFC or ACC circuits, those circuits must be specialized for dealing with interference from negative stimuli rather than emotional stimuli more broadly. Another possibility raised in the following section is that the negativity avoidance effects seen among older adults are the result of a greater chronic focus on emotion regulation goals.

## Downregulating negative affect may be more of a default mode for older adults than for younger adults

In the previous sections we reviewed evidence that, compared with younger adults, older adults favor different emotion regulation strategies and are more adept at avoiding emotional distraction. Another important factor to consider is how likely they are to focus on regulating emotion in the first place. Socioemotional selectivity theory (SST) posits that as perceived time left in life becomes more limited, people shift motivational priorities to focus more on regulating emotions (and in particular, experiencing more positive, and less negative, emotions) and less on other goals such as information seeking.[43] Indeed, in their self-reports, older adults are more likely than younger or middle-aged adults to endorse statements such as, "I try hard to stay in a neutral state and to avoid emotional situations," which indicate a focus on controlling emotions.[44] Their responses to problems presented in vignettes involve more impulse control than those of younger adults[45] and are more emotion-focused.[14,46]

In addition to age differences in self-reports about emotion regulation, in laboratory studies involving emotional stimuli, older adults often selectively attend to and remember positive stimuli compared with negative stimuli more than younger adults do.[47,48] In contrast, it seems that younger adults need reminders to get them to focus on regulating emotion—younger or middle-aged adults' memories become more consistent with emotion regulation goals (and more emotionally similar to older adults' memories) when they are reminded of their emotions by being asked to rate or think about their current emotions.[49,50] The "cognitive-control model" extension of SST posits that older adults' positivity effect results from older adults' chronic activation of emotion regulation goals.[47,51–53] Because, compared with younger adults, older adults focus more on regulating emotions when processing new stimuli, they should be more likely to engage cognitive control resources in the service of the emotional goals. Consistent with the cognitive-control model, older adults with higher cognitive control abilities are more likely to show positivity effects (i.e., relatively more attention to, and better memory for, positive compared with negative information than younger adults) than those with lower cognitive control abilities.[52,54] Furthermore, reducing the availability of cognitive resources for emotion regulation by having participants engage in a concurrent task that demands cognitive control reverses older adults' positivity effect in attention[17] and memory.[52,c]

Also consistent with the role of top–down processes, in eye-tracking studies positivity effects in attention are not evident in the first eye fixation but emerge after that. For instance, when an emotional picture is shown next to a neutral picture, both younger and older adults' first fixation tends to be on the emotional stimulus,[17] consistent with findings reviewed earlier that older adults detect emotionally arousing stimuli as effectively as younger adults. However, compared with younger adults, fewer of older adults' remaining fixations during the six-seconds display time are devoted to the negative stimuli and more are devoted to the positive stimuli (Fig. 2). Older adults also show attentional biases away from negative faces in a dot probe task,[18] and in this type of task, the attentional bias against angry faces takes about a second to emerge.[56] Furthermore, eye-tracking analyses over a 4-s duration show that older adults show the strongest attentional biases against angry faces and toward happy faces in the later portions of the trial.[57] Of particular relevance for the emotion regulation account of the positivity effect, older adults' attentional biases away from angry faces is mediated by how much they endorse everyday use of emotion suppression strategies.[56]

Emotional stimuli also influence subsequent visual processing differently for younger and older adults. An event-related potential (ERP) study in which a face was displayed for 400–800 ms and then a checkerboard pattern was flashed for 100 ms over

[c]In contrast, distraction from a secondary task that requires little cognitive control (hearing one word at the beginning of a picture presentation period and immediately indicating whether it was a word or a non-word) does not influence older adults' positivity effect in attention.[54]

the face found that younger adults showed enhanced early visual processing of the checkerboard (as indicated by greater P1 response) when the faces were angry, happy, or sad than when they were neutral.[58] Although there was no overall main effect of age on the P1 amplitude, there were age differences in the influence of emotion. Older adults showed no emotional advantages and actually showed smaller P1 amplitudes when faces were angry than when they were neutral. This suggests that older adults withdrew their attention from angry faces, reducing subsequent visual processing of probes appearing over those faces.

A study examining age differences in binocular rivalry effects for emotional faces also revealed that older adults suppress processing of angry faces and enhance processing of happy faces.[59] In this study, a picture of a house was presented to one eye and a picture of a face was presented to the participant's other eye on each trial. Under these conditions, people perceive the two images alternating, as they compete in the brain for perceptual dominance. As seen in previous studies of binocular rivalry, for younger adults both happy and angry faces dominated perception for longer durations than neutral faces. But for older adults, the angry faces were perceived for less time than the neutral faces whereas the happy faces were perceived for longer than the neutral faces. Although people have limited voluntary control over binocular rivalry dominance, paying attention to one stimulus increases its dominance.[60,61] These age differences in binocular rivalry in addition to the ERP P1 findings described earlier indicate that older adults' suppression of processing angry faces impacts even basic visual perception.

Another indication of age differences in whether positive or negative stimuli are attended more deeply come from ERP studies that assessed the late positive potential while younger and older adults studied pictures.[62,63] The late positive potential is a positive deflection in the brain electrophysiological response to a stimulus occurring about 400 ms after stimulus onset. It is greater in response to stimuli with motivational relevance than in response to neutral stimuli[64] and is increased both by automatic and directed attention.[65] Younger adults showed the largest late positive potentials to negative stimuli, whereas older adults showed either similar late positive potentials to negative and positive stimuli[62] or the greatest response to positive stimuli.[63] Older adults subsequently recalled fewer negative pictures than younger participants did, but similar numbers of positive and neutral pictures, and the difference in the ERP amplitude for positive versus negative pictures was marginally significantly related to the difference in number of positive versus negative pictures recalled.[63]

## Older adults show greater PFC activity for emotional than neutral stimuli

The studies reviewed in the previous section show that older adults tend to process positive stimuli more deeply while ignoring negative stimuli, at least when they have cognitive control resources available. If such effects result from a more chronic focus on emotion regulation among older adults than younger adults, they should be associated with additional recruitment of prefrontal control regions. Even if older adults tend to use different strategies than younger adults, PFC should still be involved.

Indeed, in fMRI studies contrasting brain activity while processing negative faces or pictures versus neutral faces or pictures, older adults tend to show more dorsolateral and ventrolateral PFC activity and less amygdala activity than younger adults.[66–73] Findings of greater PFC and reduced amygdala responses among older adults than younger adults when viewing negative stimuli may reflect more effective spontaneous emotion regulation. Consistent with this possibility, in a study in which older adults were asked to diminish their affective response to negative stimuli, those with a negative correlation between ventromedial PFC and amygdala had steeper, more normative declines in stress hormones during the day, suggesting that PFC–amygdala interactions may contribute to effective emotion and stress regulation.[74]

While older adults generally show more PFC activity for negative than neutral pictures or faces in studies that involve passive viewing or simple ratings of the pictures, when tasks instruct participants to deeply process presented stimuli, older adults show a larger increase in PFC activity in response to positive stimuli than negative stimuli compared with younger adults (Fig. 4).[75,76] This suggests that PFC control regions may be recruited either to more deeply process positive stimuli or to diminish emotional responses to negative stimuli, and that recruitment for one or the other purpose varies depending on the task context.

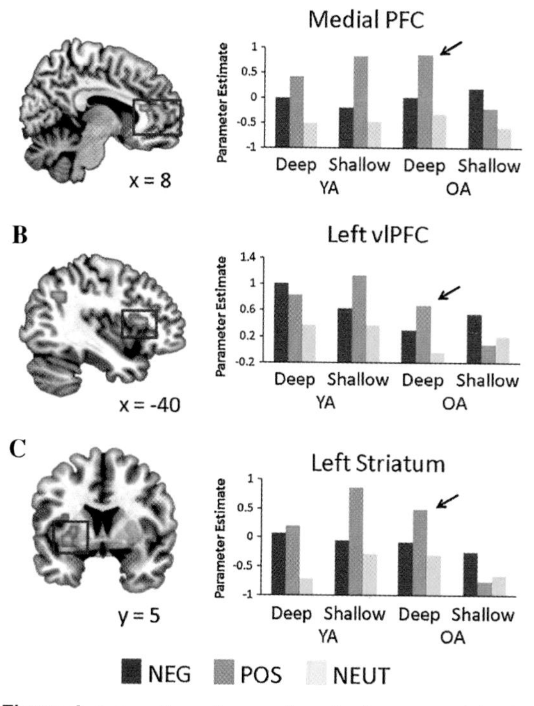


**Figure 4.** Interactions of age, task, and valence on activity in (A) medial prefrontal cortex, (B) left ventrolateral prefrontal cortex, and (C) striatum, all indicating a greater difference between positive than negative trials during deep versus shallow processing in older adults. Figure is from Ritchey *et al.*[76]

As reviewed earlier, ventromedial PFC maintains its cortical thickness in normal aging, whereas lateral and dorsal regions show clear decline (Fig. 1). This raises the question of whether ventromedial PFC plays a more important role in emotion regulation for older adults than for younger adults. As shown in Figure 5A, older adults' greater activation in PFC when processing negative stimuli does not tend to occur in ventromedial PFC. Instead, the activations tend to be in more dorsal regions of PFC. While the peak regions activated more for older adults for positive than negative stimuli also extend into dorsal PFC, they show more overlap with ventromedial PFC. Thus, when compared with younger adults, older adults' greater PFC activity for negative than neutral stimuli is not centered in ventromedial PFC, whereas their greater PFC activity for positive than negative stimuli is more likely to be seen in ventromedial PFC (Fig. 5B).

Whether or not people remember information can help indicate how deeply they processed it. But there are many different types of processing that can enhance later memory for new information. For instance, when seeing a picture of a forest scene, someone might focus on trying to identify the types of tree shown, whereas someone else might think about how the picture makes them feel happy and relaxed. Both types of associations should increase the likelihood of remembering the picture, but be associated with different patterns of brain activity during encoding. Interestingly, a number of studies have revealed age differences in the patterns of brain activity associated with successful encoding of emotional stimuli. For instance, in a study in which participants rated whether objects would fit in a file drawer, there was an interaction of item valence and age group on later recognition, such that older adults were as effective as younger adults at remembering positive items but less effective at remembering negative and neutral items.[77,78] Successful recognition of positive items was more associated with ventromedial PFC for older adults than younger adults.[78] Also, during encoding of positive information, the ventromedial PFC and amygdala were more strongly correlated with hippocampal activity for older adults than for younger adults.[77] Medial PFC is activated by thinking about ones's relationship to external stimuli,[79] and so one potential explanation for older adults' greater ventromedial PFC activity while processing positive stimuli is that they are thinking more about how the positive stimuli relate to themselves than younger adults do.[80]

Age differences have also been found in brain regions associated with encoding of fearful faces,[81] as successful encoding of fearful faces was more associated with amygdala and hippocampus in younger adults than older adults and more associated with insula and dorsolateral PFC in older adults than younger adults.

In another study,[82] brain activity during memory encoding of negative stimuli differed by age group, with older adults showing less functional connectivity than younger adults between the amygdala and hippocampus but more functional connectivity between the amygdala and dorsolateral PFC. As expected based on previous studies, younger adults showed a larger advantage in memory for negative stimuli than older adults did. Thus, across studies, there are contrasting patterns of older adults' amygdala connectivity during memory encoding for

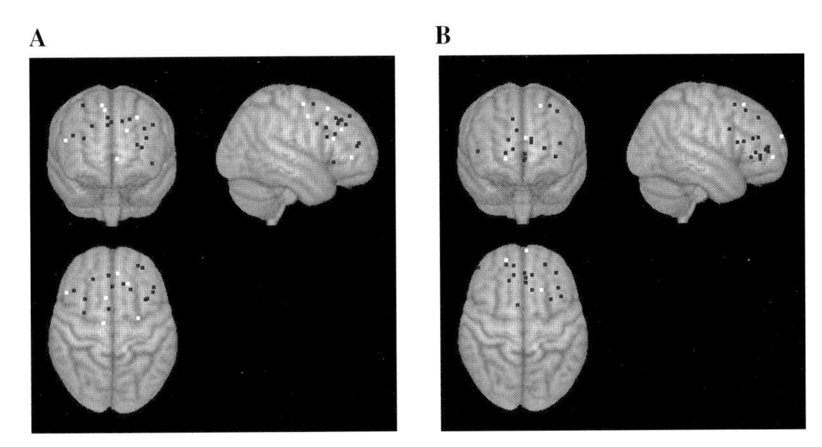


**Figure 5.** Across fMRI studies reviewed by Nashiro, Sakaki, and Mather,[53] peak activations are shown for age differences in prefrontal involvement while processing (A) negative stimuli and (B) positive stimuli. Peak activations that were greater for older than younger are shown in black; those greater for younger are shown in white. On average, the contrasts in which older adults showed greater activity than younger adults for negative stimuli tended to yield more dorsal peak activity foci (mean MNI $Z = 25.7$) than the contrasts in which older adults showed greater activity than younger adults for positive stimuli (mean MNI $Z = 18.5$; comparison of negative and positive $t(29) = 2.33$, $P < .05$). [Correction added after online publication. 03/26/2012: the previous symbol was changed from > to <.] Figures are from Nashiro *et al.* [53]

positive stimuli[77] (ventromedial PFC) and negative stimuli[82] (dorsolateral PFC).

Elucidating how the PFC–amygdala connectivity changes with age is important, as preservation of this connectivity is associated with well-being. A study in which older adult participants (aged 64–89) simply viewed a series of positive, negative, and neutral pictures while in an fMRI scanner reveals that PFC–amygdala connectivity during memory encoding relates to well-being in everyday life.[83] Participants with high life satisfaction showed stronger correlations in activity between the orbitofrontal cortex, ventromedial PFC, hippocampus, amygdala, and thalamus during viewing positive pictures than during viewing negative pictures, whereas these differences in functional connectivity were weaker for those lower in life satisfaction. Thus, emotion-processing and memory-encoding regions showed more coordinated activity during viewing positive than negative pictures for those high in life satisfaction.

The findings reviewed in this section reveal that, compared with younger adults, older adults show a greater increase in PFC and decrease in amygdala activity when just asked to view negative stimuli compared with neutral stimuli, but that when asked to deeply process stimuli, they show a greater increase in PFC activity for positive over negative stimuli.[53] In these contexts in which there are no instructions to regulate emotions, older adults tend to engage more dorsal PFC cognitive control regions (possibly to diminish the emotional impact of negative stimuli) and more ventromedial PFC (possibly to engage more deeply with self-relevant aspects of positive stimuli). Furthermore, those who show more coordination of PFC regions with memory encoding regions when processing emotional information appear to be more likely to have high levels of well-being in everyday life.

## When instructed to regulate, age differences depend on type of strategy and the specific instructions given

The findings reviewed in the two previous sections suggest that older adults tend to engage in emotion regulation as a default mode whereas younger adults need to be reminded of their emotions to activate emotion regulation goals. They also indicate that older adults select different emotion regulation strategies than younger adults. However, they do not address the question of whether older adults are generally better than younger adults at regulating emotion when emotion regulation is the top priority for both age groups.

Laboratory studies that instruct people to up- or downregulate their emotions using a specific strategy while viewing emotional film clips or pictures reveal different patterns of findings across types of regulation strategies. One consistent pattern is that older adults are not impaired at suppressing the expression of emotion when instructed to do so.[84–86] Furthermore, older adults also show less cost of suppressing emotional expression in terms of memory for the stimuli shown during suppression trials.[87] The findings of a reduced cost of suppression for older adults suggest that they use different strategies than younger adults when trying to suppress their outward emotional expressions, but that these strategies do not involve ignoring the emotion elicitor.

In contrast, studies that instruct participants to reappraise emotional stimuli reveal mixed findings, which may result from the different reappraisal instructions given. For instance, when attention deployment is controlled so that participants must attend to an aversive stimulus, older adults are worse than younger adults at diminishing their negative reactions by reappraising the meaning of those stimuli.[88] However, when attention deployment is not controlled, older adults are better than younger adults at diminishing their negative reactions when given reappraisal instructions.[89] One possibility is that, when attention deployment is not controlled, older adults will try to ignore the negative stimuli rather than reappraise them, despite the instructions.

Other studies suggest that older adults are better than younger adults at implementing some types of refocusing or reappraisal strategies (e.g., focusing away from a negative stimulus by thinking about a positive memory[86]) but worse at others (e.g., thinking of themselves as an emotionally detached and objective third party[90]).

These differences across studies suggest that focusing on nonemotional aspects of a negative effect-inducing situation is not an effective strategy for older adults, whereas focusing on something else works well for them. This may relate to age-related increases in distractibility;[27] ignoring one aspect of a situation while attending to another may be more challenging than ignoring the whole situation while attending to something unrelated.

In the study that involved a detachment reappraisal strategy,[90] participants were scanned during the task and, in general, younger and older adults both showed greater PFC but less amygdala activity during "reappraise" trials than during "experience" trials. Thus, when people are trying to regulate emotion using a detached reappraisal strategy, younger and older adults show similar patterns of brain activity. However, one age difference that emerged was that for the negative reappraise versus experience contrast, younger adults showed a larger difference in activity in the left lateral inferior frontal gyrus than older adults did. For older adults, but not younger adults, activity in this region was related to regulation success on negative trials. That age differences emerged in the left inferior frontal cortex is intriguing given that this region may be recruited to help avoid negative distraction as discussed earlier. Across all participants in this study (controlling for age), higher scores on a composite score from a battery of cognitive tests testing response speed, memory, and executive function predicted less amygdala activation when down-regulating negative affect. Thus, higher cognitive function was associated with more effective emotion regulation when instructed to do so, both for younger and for older adults.

More striking age differences in PFC activity were seen in another fMRI reappraisal study in which gaze was controlled while participants reappraised the meaning of negative pictures.[88] In this study, older adults showed less ventrolateral and dorsomedial PFC activity than younger adults while reappraising negative images. Thus, it seems that the less that older adults are able to use attention redeployment strategies, the more that age differences emerge.

Another interesting age difference is that, when asked to downregulate negative reactions after watching a disgusting video, older adults showed less cost in terms of their concurrent working memory performance than did younger adults.[91] Participants were told to use whatever emotion regulation strategy they had available, thus it could be that the emotion regulation strategies employed by older adults were less cognitively taxing than the strategies employed by younger adults.

In summary, previous studies comparing younger and older adults' effectiveness at implementing specific emotion regulation strategies when instructed to do so reveal mixed findings.[92] Older adults are as good as younger adults at suppressing facial expression of their emotions and experience less cost of this strategy in terms of what they remember. Reappraisal or refocusing success varies widely, with some studies finding that younger adults do better and other studies finding that older adults do better. These differences likely stem from differences in reappraisal instructions. In particular, older adults do well when cued to direct their attention to something else entirely[86] or to reappraise the meaning of negative stimuli,[89] but not when they are asked to be emotionally detached[90] or to focus on nonemotional aspects of the emotion-eliciting event.[89]

At the outset of this paper, I noted that there are two obvious potential explanations for the paradox of emotional well-being in aging. The first is that older adults are better at regulating their emotions. However, as should now be clear, neither younger nor older adults are consistently better than the other group at regulating emotion. Instead, the story is more nuanced. Older adults excel at some modes of regulating emotion (such as avoiding attending to negative stimuli) but not at others (such as certain types of reappraisal). In addition, the research reviewed here suggests that, in their everyday lives, older adults favor different emotion regulation strategies and focus more of their cognitive resources on regulating emotion than younger adults do. These age-related shifts in strategy selection and resource allocation may shape well-being more than age-related changes in emotion regulation skill level.

## Age differences in the initial emotional sensation

The other potential explanation for the paradox of emotional well-being in aging mentioned at the outset of this paper was that negative stimuli and emotions may be less potent for older adults than for younger adults, and therefore older adults experience less negative affect. There are at least a couple of ways that negative stimuli might feel less potent for older adults. The first is due to declines in interoceptive processes. Part of what determines initial emotional sensations are cues from brain regions that perceive body sensations, such as heartbeats, breath, gastrointestinal sensations, and flushing of the skin. These contribute to people's interoception of their own body states and help shape the feeling and intensity of emotions.[93,94] With age, people are less aware of visceral sensations such as esophageal pain,[95] rectal distension,[96] gastric distension,[97] and heartbeats.[98] Multiple factors may contribute to this reduced perception of bodily states. Brain regions critical for interoception, such as the insula, decline in volume with age[99] and the specificity of neural representations decreases in the somatosensory cortex.[100] Changes in signal transmission in afferent fibers and changes in neurotransmitter quantity and metabolism may also contribute to the age-related declines in interoception.[101,102] Unfortunately, existing research provides little information about how age differences in interoception may relate to emotional experience and general well-being, but it is an important question for future research.

The second way that negative stimuli might feel less potent is if the amygdala's preferential processing of negative arousing stimuli declines with age. This is the explanation for the positivity effect put forward in the "aging-brain model,"[103] in which age-related declines lead the amygdala to activate less in response to negative stimuli, which in turn diminishes the arousal response to those stimuli and impairs later memory for them.

Certainly, as already reviewed, compared with younger adults, older adults do typically show less amygdala activity in response to negative stimuli.[67,68,71,104] However, reduced amygdala response is not necessarily an indication of poor function. Studies of emotion regulation indicate that people show less amygdala response to negative stimuli when they are asked to diminish their emotional response to those stimuli along with greater prefrontal activity,[22,105] and so older adults' pattern of greater PFC and reduced amygdala activity when told to observe negative stimuli may reflect spontaneous (uninstructed by the experimenter) emotion regulation. Furthermore, the amygdala is one of the regions showing the most declines in volume in Alzheimer's disease,[106] yet patients with Alzheimer's disease show hyperactive amygdala responses to novel fearful faces relative to healthy older adults.[107] Thus, diminished responses might not be the pattern expected from an amygdala in decline.

Other data also suggest that age differences in amygdala function reflect shifts in processing strategies more than declines in functional capacity. For instance, fMRI studies that included positive as well as negative and neutral pictures reveal an age by valence interaction in the amygdala (e.g., Fig. 6) such that, unlike younger adults, older adults showed the most amygdala activity in response to positive rather than negative pictures.[69,104] (See also Ref. 83.) Thus, older adults' decreased amygdala response is selective to negative stimuli and does not extend to positive stimuli—a pattern indicating that the amygdala maintains the capacity to respond to emotionally arousing stimuli.

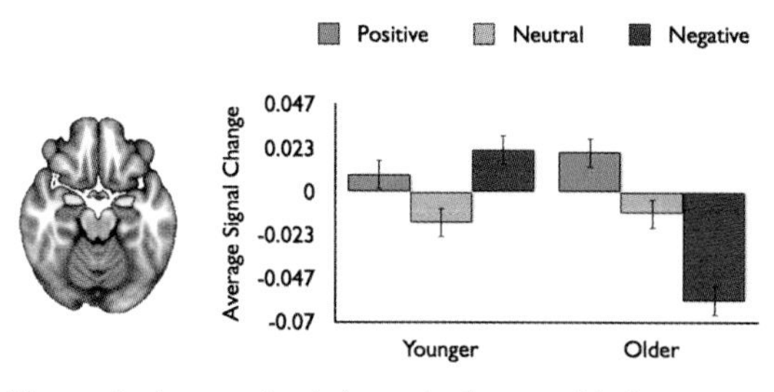


**Figure 6.** Average signal change in the amygdala for younger and older adults while viewing positive, neutral, or negative pictures. Figure is adapted from Mather *et al.*[104]

Another interesting pattern across studies noted by Kensinger and Leclerc[80] is that older adults showed as much of an increase as younger adults in amygdala activity in response to novel fearful compared with familiar neutral faces in studies in which faces were presented rapidly (at a 200-ms rate),[108,107] but showed less amygdala response to negative stimuli than younger adults in studies in which faces were presented more slowly.[67,68,71] Thus, age-related declines in negative affect are unlikely the result of impairments in amygdala initial responsiveness to negative stimuli, but rather arise because of how older adults subsequently respond.

Behavioral data also indicate that younger and older adults show similar responses to arousing negative stimuli initially, but then diverge. For instance, older adults show at least as much of an advantage as younger adults in detecting emotional or threatening stimuli.[109,110] Indeed, within one study, older adults showed as much of an advantage as younger adults in detecting angry faces in a matrix of happy or neutral faces but were more efficient than younger adults at detecting happy or neutral faces in a matrix of angry faces, indicating they were less distracted by nontarget angry faces than other types of nontarget faces.[111] Thus, older adults' enhanced ability to avoid distraction from negative stimuli does not stem from impairments in being able to detect those stimuli.

These brain and behavioral data showing similar rapid responses to potentially threatening stimuli among younger and older adults argue against the idea that older adults' shifts away from attending to negative stimuli are due to declines in amygdala function and the ability to monitor and quickly notice negative stimuli. Instead, the age differences emerge in later processing, when strategic processes have more influence.

In summary, there are age-related declines in the types of interoceptive sensation that contribute to emotional experience. Such declines may decrease the intensity of some negative emotions, making the task of regulating emotion less challenging for older adults than for younger adults. However, another aspect of initial emotional experience—detecting negative stimuli—still seems intact in normal aging and the amygdala still responds to emotionally arousing stimuli that are positive or that are negative but displayed too rapidly for strategic processes to diminish responding.

## Late-life depression is often associated with PFC damage

So far, this review has focused on the general pattern of well-maintained emotional well-being in late life. However, some people do get unhappier with age, and some even experience an episode of major depression for the first time late in life. In this section, I review how depression in late life might relate to the neural systems already discussed.

In the U.S. population, the incidence of depression in the past 12 months decreases across age cohorts, with the highest incidence among those 18–29 and the lowest among those greater than 65.[112] In addition, the likelihood of experiencing one's first episode of major depression is lower among those older than 65 than in the other age groups. However, among those older adults experiencing major depression, it appears that for at least half of them, this is their first episode of depression.[113]

Late-onset depression differs in many ways from early-onset depression. It seems to be less genetically determined[114] and is associated with frontostriatal abnormalities, executive control deficits, and the presence of vascular disease. This syndrome is known as "vascular depression," and the hypothesis is that cerebrovascular disease leads to damage in white matter tracts connecting frontal regions with striatum, amygdala, and hippocampus.[115–118] According to the vascular depression hypothesis, this damage impairs basic emotion regulation functions and leads to depression that is less amenable to antidepressant treatment. For instance, patients over the age of 60 who were still depressed after a 12-week treatment with escitalopram, a selective serotonin reuptake inhibitor antidepressant, initially had more white matter abnormalities in multiple frontal brain regions than patients who responded to the antidepressants,[119,120] and nonresponders also initially had lower semantic fluency performance than responders.[121] Also, suggesting a link between PFC function and vascular depression are findings that repetitive transcranial magnetic stimulation to the left dorsolateral PFC led to greater remission from depression among patients with clinically defined vascular depression than did sham stimulation.[122]

Analysis of a large sample of older depressed patients indicated that a vascular depression profile could be distinguished from other types of depression by a high probability of having deep white matter MRI hyperintensities, executive dysfunction, and late-life onset of the depression.[123] This subtype was nearly perfectly identified by the presence of deep white matter hyperintensities, although including executive dysfunction as a measure increased the accuracy of class membership identification. In addition, a meta-analysis of 30 studies revealed that late-life onset of depression was associated with more frequent and intense white matter abnormalities than earlier-onset depression.[124]

Impaired ACC function has been identified as a key feature of depression in younger adults,[125] which may be due to the role of the ACC in integrating ventral and subcortical networks with dorsal cognitive networks in ways that support the experience and regulation of emotion.[126] Older depressed patients who did not respond to antidepressant treatment had smaller dorsal and rostral ACC volume than those who did remit.[127] Thus, in older adults, preservation of the ACC may be an important determinant in recovery from depression.

In summary, depression in later life is often associated with white matter damage in frontostriatal and limbic regions and with decline in executive function. These associations suggest that age-related vulnerability to PFC decline, especially in PFC white matter, affects emotion regulation abilities. Older adults who, because of vascular disease or other risk factors, show the most decline in executive function, also are likely to be those who lose the ability to regulate emotion successfully and thus become more vulnerable to depression.

## Conclusions

Knowing about the declines expected in cognitive function and physical health for older adults might lead one to predict significant declines in well-being and emotion regulation. But a paradox of aging is that the frequency of negative affect and incidence of depression actually decrease. What might explain this surprising pattern? This review highlights the fact that age-related changes in emotion cannot be explained simply by which brain regions decline most structurally with age, although such brain changes may help shape emotion regulation strategies. In addition, other obvious explanations, such as the ideas that older adults are more skilled at regulating emotions or experience negative stimuli less intensely, also are insufficient to explain the full pattern of age-related changes. Instead, it is important to consider age-related shifts in preferred strategies and priorities and how these interact with the strengths and vulnerabilities of the aging brain to better understand how emotional experience changes with age.

Compared with younger adults, older adults pay attention to different types of emotional stimuli and prefer to regulate their emotions in different ways. In their attention and memory, older adults often show a positivity effect, in which they favor positive relative to negative stimuli more than younger adults do. Despite age impairments in ignoring neutral distractions, older adults are even better than younger adults at ignoring distracting negative stimuli. Furthermore, with age preferred emotion regulation strategies shift to favor suppression over rumination or reappraisal. When viewing emotional stimuli, older adults often show an increase in prefrontal activity (relative to viewing neutral stimuli) that is greater than that shown by younger adults.

How do these findings relate to the fact that that ventromedial prefrontal brain regions, supporting the types of emotion processing strategies older adults excel at and prefer, decline less structurally than dorsal and lateral prefrontal brain regions (Fig. 1)? Ventromedial PFC has been linked to many aspects of emotion regulation, and older adults have been shown to engage it more when deeply processing positive stimuli than younger adults. However, not all of the age changes in emotion processing can be explained by a greater reliance on ventromedial PFC among older adults. In younger adults, lateral left inferior frontal gyrus is associated with ignoring negative distraction, which is something that older adults excel at despite significant age-related declines in cortical thickness in this region. Older adults often show greater increases in dorsal PFC activity when viewing negative stimuli (relative to neutral stimuli) compared with younger adults (Fig. 5A), again despite the fact that there are age-related declines in this region. Thus, older adults appear to recruit a variety of prefrontal regions when confronted with negative stimuli—including prefrontal subregions that show significant age-related structural decline as well as those that do not.

This pattern of findings can be explained by the idea that, in addition to age-related changes in brain circuitry underlying emotional processing, there are age-related changes in emotional goals that lead older adults to focus more on regulating emotions. Older adults therefore are more likely to recruit prefrontal resources to regulate emotions than younger adults are. Some emotion regulation strategies, such as reappraisal, may be more difficult for older than younger adults to successfully implement. But unless specifically asked to use such strategies, older adults may be able to downregulate negative affect more effectively than younger adults by focusing more of their resources on this goal and by using strategies that work well for them, such as selectively attending to or ignoring stimuli.

In conclusion, the brain shows many changes with age, and in cognition, these changes mostly result in declines in function. This is not the case for emotional processing. Understanding how some emotional functions can be well-maintained despite declines in brain function should not only advance theories about emotion, but should also help clinicians design interventions to support effective emotional processes throughout life.

## Acknowledgments

I would like to thank Sarah Barber, David Clewett, Kaoru Nashiro, and Alexandra Ycaza for their comments on previous versions of this review. Preparation of this paper was supported by Grants RO1AG025340 and K02AG032309 from the National Institute on Aging.

## Conflicts of interest

The author declares no conflicts of interest.

## References

1. Charles, S.T. 2010. Strength and vulnerability integration: a model of emotional well-being across adulthood. *Psychol. Bull.* **136:** 1068.
2. Carstensen, L.L., M. Pasupathi, U. Mayr & J.R. Nesselroade. 2000. Emotional experience in everyday life across the adult life span. *J. Pers. Soc. Psychol.* **79:** 644.
3. Hay, E.L. & M. Diehl. 2011. Emotion complexity and emotion regulation across adulthood. *Eur. J. Ageing* **8:** 157.
4. Neupert, S.D., D.M. Almeida & S.T. Charles. 2007. Age differences in reactivity to daily stressors: the role of personal control. *J. Gerontol. B-Psychol. Sci. Soc. Sci.* **62:** P216.
5. Birditt, K.S. & K.L. Fingerman. 2005. Do we get better at picking our battles? Age group differences in descriptions of behavioral reactions to interpersonal tensions. *J. Gerontol. B Psychol. Sci. Soc. Sci.* **60B:** P121.
6. Birditt, K.S., K.L. Fingerman & D.M. Almeida. 2005. Age differences in exposure and reactions to interpersonal tensions: a daily diary study. *Psychol. Aging* **20:** 330.
7. Anderson, S.W., J. Barrash, A. Bechara & D. Tranel. 2006 Impairments of emotion and real-world complex behavior following childhood- or adult-onset damage to ventromedial prefrontal cortex. *J. Int. Neuropsychol. Soc.* **12:** 224.
8. Allman, J. M. *et al.* 2001. The anterior cingulate cortex. *Ann. N.Y. Acad. Sci.* **935:** 107.
9. Shaw, P. *et al.* 2008. Neurodevelopmental trajectories of the human cerebral cortex. *J. Neurosci.* **28:** 3586.
10. Fjell, A. M. *et al.* 2009. High consistency of regional cortical thinning in aging across multiple samples. *Cereb. Cortex* **19:** 2001.
11. Nolen-Hoeksema, S. & A. Aldao. 2011. Gender and age differences in emotion regulation strategies and their relationship to depressive symptoms. *Pers. Indv. Diffs.* **51:** 704.
12. Marquez-Gonzalez, M., M.I.F. de Troconiz, I.M. Cerrato & A.L. Baltar. 2008. Emotional experience and regulation across the adult lifespan: comparative analysis in three age groups. *Psicothema* **20:** 616.
13. Yeung, D.Y., C.K.M. Wong & D.P.P. Lok. 2011. Emotion regulation mediates age differences in emotions. *Aging Ment. Health* **15:** 414.
14. Blanchard-Fields, F., C. Camp & H. Casper Jahnke. 1995. Age differences in problem-solving style: the role of emotional salience. *Psychol. Aging* **10:** 173.
15. Isaacowitz, D.M., H.A. Wadlinger, D. Goren & H.R. Wilson. 2006. Selective preference in visual fixation away from negative images in old age? An eye tracking study. *Psychol. Aging* **21:** 40.
16. Isaacowitz, D.M., H.A. Wadlinger, D. Goren & H.R. Wilson. 2006. Is there an age-related positivity effect in visual attention? A comparison of two methodologies. *Emotion* **6:** 511.
17. Knight, M. *et al.* 2007. Aging and goal-directed emotional attention: distraction reverses emotional biases. *Emotion* **7:** 705.
18. Mather, M. & L.L. Carstensen. 2003. Aging and attentional biases for emotional faces. *Psychol. Sci.* **14:** 409.
19. Mather, M. 2010. Aging and cognition. *Wiley Interdiscip. Rev. Cogn. Sci.* **1:** 346.
20. Isaacowitz, D.M., K. Toner, D. Goren & H.R. Wilson. 2008. Looking while unhappy: mood-congruent gaze in young adults, positive gaze in older adults. *Psychol. Sci.* **19:** 848.
21. Urry, H.L. & J.J. Gross. 2010. Emotion regulation in older age. *Curr. Dir. Psychol. Sci.* **19:** 352.
22. Goldin, P.R., K. McRae, W. Ramel & J.J. Gross. 2008. The neural bases of emotion regulation: reappraisal and suppression of negative emotion. *Biol. Psychiatry* **63:** 577.

23. Gyurak, A., J.J. Gross & A. Etkin. 2011. Explicit and implicit emotion regulation: a dual-process framework. *Cogn. Emot.* **25:** 400.
24. Phillips, M.L., W.C. Drevets, S.L. Rauch & R. Lane. 2003. Neurobiology of emotion perception I: the neural basis of normal emotion perception. *Biol. Psychiatry* **54:** 504.
25. Ochsner, K.N. & J.J. Gross. 2008. Cognitive emotion regulation: insights from social cognitive and affective neuroscience. *Curr. Dir. Psychol. Sci.* **17:** 153.
26. Guerreiro, M.J.S., D.R. Murphy & P.W.M. Van Gerven. 2010. The role of sensory modality in age-related distraction: a critical review and a renewed view. *Psychol. Bull.* **136:** 975.
27. Healey, M.K., K.L. Campbell & L. Hasher. 2008. Cognitive aging and increased distractibility: costs and potential benefits. In *Prog. Brain Res.* Vol. 169. W. S. Sossin, J.-C. Lacaille, V. F. Castelluci & S. Belleville, Eds.: 353. Elsevier. Amsterdam.
28. LaMonica, H.M. *et al.* 2010. Differential effects of emotional information on interference task performance across the life span *Front. Aging Neurosci.* **141**.
29. Wurm, L.H. *et al.* 2004. Performance in auditory and visual emotional Stroop tasks: a comparison of older and younger adults. *Psychol. Aging* **19:** 523.
30. Ashley, V. & D. Swick. 2009. Consequences of emotional stimuli: age differences on pure and mixed blocks of the emotional Stroop. *Behav. Brain Funct.* **5:** 14.
31. Monti, J.M., S. Weintraub & T. Egner. 2010. Differential age-related decline in conflict-driven task-set shielding from emotional versus non-emotional distracters. *Neuropsychologia* **48:** 1697.
32. Ebner, N.C. & M.K. Johnson. 2010. Age-group differences in interference from young and older emotional faces. *Cogn. Emot.* **24:** 1095.
33. Thomas, R.C. & L. Hasher. 2006. The influence of emotional valence on age differences in early processing and memory. *Psychol. Aging* **21:** 821.
34. Stawski, R.S., J. Mogle & M.J. Sliwinski. 2011. Intraindividual coupling of daily stressors and cognitive interference in old age. *J. Gerontol. B Psychol. Sci. Soc. Sci.* **66:** 121.
35. Dolcos, F., P. Kragel, L.H. Wang & G. McCarthy. 2006. Role of the inferior frontal cortex in coping with distracting emotions. *NeuroReport* **17:** 1591.
36. Dolcos, F. & G. McCarthy. 2006. Brain systems mediating cognitive interference by emotional distraction. *J. Neurosci.* **26:** 2072.
37. Shafritz, K.M., S.H. Collins & H.P. Blumberg. 2006. The interaction of emotional and cognitive neural systems in emotionally guided response inhibition. *NeuroImage* **31:** 468.
38. Driscoll, D.M. 2009. The effects of prefrontal cortex damage on the regulation of emotion, Dissertation, University of Iowa, Iowa City, IA.
39. Egner, T., A. Etkin, S. Gale & J. Hirsch. 2008. Dissociable neural systems resolve conflict from emotional versus nonemotional distracters. *Cereb. Cortex* **18:** 1475.
40. Whalen, P.J. *et al.* 1998. The emotional counting Stroop paradigm: a functional magnetic resonance imaging probe of the anterior cingulate affective division. *Biol. Psychiatry* **44:** 1219.
41. Samanez-Larkin, G.R. *et al.* 2009. Selective attention to emotion in the aging brain. *Psychol. Aging* **24:** 519.
42. Brassen, S., M. Gamer & C. Buchel. 2011. Anterior cingulate activation is related to a positivity bias and emotional stability in successful aging. *Biol. Psychiatry* **70:** 131.
43. Carstensen, L.L., D.M. Isaacowitz & S.T. Charles. 1999. Taking time seriously: a theory of socioemotional selectivity. *Am. Psychol.* **54:** 165.
44. Lawton, M.P., M.H. Kleban, D. Rajagopal & J. Dean. 1992. Dimensions of affective experience in three age groups. *Psychol. Aging* **7:** 171.
45. Diehl, M., N. Coyle & G. Labouvie-Vief. 1996. Age and sex differences in strategies of coping and defense across the life span. *Psychol. Aging* **11:** 127.
46. Blanchard-Fields, F. & C.J. Camp. 1990. Affect, individual differences, and real world problem solving across the adult life span. In *Aging and Cognition: Knowledge Organization and Utilization*. T. Hess, Ed.: 461. North-Holland. Oxford, England.
47. Mather, M. & L.L. Carstensen. 2005. Aging and motivated cognition: the positivity effect in attention and memory. *Trends Cogn. Sci.* **9:** 496.
48. Scheibe, S. & L.L. Carstensen. 2010. Emotional aging: recent findings and future trends. *J. Gerontol. B Psychol. Sci. Soc. Sci.* **65:** 135.
49. Kennedy, Q., M. Mather & L.L. Carstensen. 2004. The role of motivation in the age-related positivity effect in autobiographical memory. *Psychol. Sci.* **15:** 208.
50. Mather, M. & M.K. Johnson. 2000. Choice-supportive source monitoring: do our decisions seem better to us as we age? *Psychol. Aging* **15:** 596.
51. Kryla-Lighthall, N. & M. Mather. 2009. The role of cognitive control in older adults' emotional well-being. In *Handbook of Theories of Aging*. V. Berngtson, D. Gans, N. Putney & M. Silverstein, Eds.: 323. Springer Publishing. New York.
52. Mather, M. & M. Knight. 2005. Goal-directed memory: the role of cognitive control in older adults' emotional memory. *Psychol. Aging* **20:** 554.
53. Nashiro, K. *et al.* 2011. Age differences in brain activity during emotion processing: reflections of age-related decline or increased emotion regulation? *Gerontology*. In press.
54. Petrican, R., M. Moscovitch & U. Schimmack. 2008. Cognitive resources, valence, and memory retrieval of emotional events in older adults. *Psychol. Aging* **23:** 585.
55. Allard, E.S. & D.M. Isaacowitz. 2008. Are preferences in emotional processing affected by distraction? Examining the age-related positivity effect in visual fixation within a dual-task paradigm. *Aging Neuropsychol. Cogn.* **15:** 725.
56. Orgeta, V. 2011. Avoiding threat in late adulthood: testing two life span theories of emotion. *Exp. Aging Res.* **37:** 449.
57. Isaacowitz, D.M., E.S. Allard, N.A. Murphy & M. Schlangel. 2009. The time course of age-related preferences toward positive and negative stimuli. *J. Gerontol. B Psychol. Sci. Soc. Sci.* **64B:** 188.
58. Mienaltowski, A. *et al.* 2011. Anger management: age differences in emotional modulation of visual processing. *Psychol. Aging* **26:** 224.
59. Bannerman, R.L., P. Regener & A. Sahraie. 2011. Binocular rivalry: a window into emotional processing in aging. *Psychol. Aging* **26:** 372.

60. Chong, S.C. & R. Blake. 2006. Exogenous attention and endogenous attention influence initial dominance in binocular rivalry. *Vision Res.* **46:** 1794.
61. Mitchell, J.F., G.R. Stoner & J.H. Reynolds. 2004. Object-based attention determines dominance in binocular rivalry. *Nature* **429:** 410.
62. Kisley, M.A., S. Wood & C.L. Burrows. 2007. Looking at the sunny side of life: age-related change in an event-related potential measure of the negativity bias. *Psychol. Sci.* **18:** 838.
63. Langeslag, S.J.E. & J.W. van Strien. 2009. Aging and emotional memory: the co-occurence of neurophysiological and behavioral positivity effects. *Emotion* **9:** 369.
64. Schupp, H.T. *et al.* 2000. Affective picture processing: the late positive potential is modulated by motivational relevance. *Psychophysiology* **37:** 257.
65. Hajcak, G., J.P. Dunning & D. Foti. 2009. Motivated and controlled attention to emotion: time-course of the late positive potential. *Clin. Neurophysiol.* **120:** 505.
66. Fischer, H. *et al.* 2005. Age-differential patterns of brain activation during perception of angry faces. *Neurosci. Lett.* **386:** 99.
67. Gunning-Dixon, F.M. *et al.* 2003. Age-related differences in brain activation during emotional face processing. *Neurobiol. Aging* **24:** 285.
68. Iidaka, T. *et al.* 2002. Age-related differences in the medial temporal lobe responses to emotional faces as revealed by fMRI. *Hippocampus* **12:** 352.
69. Leclerc, C.M. & E.A. Kensinger. 2011. Neural processing of emotional pictures and words: a comparison of young and older adults. *Dev. Neuropsychol.* **36:** 519.
70. Murty, V.P. *et al.* 2009. Age-related alterations in simple declarative memory and the effect of negative stimulus valence. *J. Cogn. Neurosci.* **21:** 1920.
71. Tessitore, A. *et al.* 2005. Functional changes in the activity of brain regions underlying emotion processing in the elderly. *Psychiatry Res.* **139:** 9.
72. St. Jacques, P.L., B. Bessette-Symons & R. Cabeza. 2009. Functional neuroimaging studies of aging and emotion: fronto-amygdalar differences during emotional perception and episodic memory. *J. Int. Neuropsychol. Soc.* **15:** 819.
73. Williams, L.M. *et al.* 2006. The mellow years? Neural basis of improving emotional stability over age. *J. Neurosci.* **26:** 6422.
74. Urry, H.L. *et al.* 2006. Amygdala and ventromedial prefrontal cortex are inversely coupled during regulation of negative affect and predict the diurnal pattern of cortisol secretion among older adults. *J. Neurosci.* **26:** 4415.
75. Gutchess, A.H., E.A. Kensinger & D.L. Schacter. 2007. Aging, self-referencing, and medial prefrontal cortex. *Soc. Neurosci.* **2:** 117.
76. Ritchey, M., B. Bessette-Symons, S.M. Hayes & R. Cabeza. 2011. Emotion processing in the aging brain is modulated by semantic elaboration. *Neuropsychologia* **49:** 640.
77. Addis, D.R., C.M. Leclerc, K.A. Muscatell & E.A. Kensinger. 2010. There are age-related changes in neural connectivity during the encoding of positive, but not negative, information. *Cortex* **46:** 425.
78. Kensinger, E.A. & D.L. Schacter. 2008. Neural processes supporting young and older adults' emotional memories. *J. Cogn. Neurosci.* **20:** 1161.
79. Northoff, G. & F. Bermpohl. 2004. Cortical midline structures and the self. *Trends Cogn. Sci.* **8:** 102.
80. E.A. Kensinger & C.M. Leclerc. 2009. Age-related changes in the neural mechanisms supporting emotion processing and emotional memory. *Eur. J. Cogn. Psychol.* **21:** 192.
81. Fischer, H., L. Nyberg & L. Backman. 2010. Age-related differences in brain regions supporting successful encoding of emotional faces. *Cortex* **46:** 490.
82. St. Jacques, P.L., F. Dolcos & R. Cabeza. 2009. Effects of aging on functional connectivity of the amygdala for subsequent memory of negative pictures: a network analysis of functional magnetic resonance imaging data. *Psychol. Sci.* **20:** 74.
83. Waldinger, R.J., E.A. Kensinger & M.S. Schulz. 2011. Neural activity, neural connectivity, and the processing of emotionally valenced information in older adults: links with life satisfaction. *Cogn. Affect. Behav. Neurosci.* **11:** 426.
84. Kunzmann, U., C.S. Kupperbusch & R.W. Levenson. 2005.Behavioral inhibition and amplification during emotional arousal: a comparison of two age groups. *Psychol. Aging* **20:** 144.
85. Magai, C. *et al.* 2006. Emotion experience and expression across the adult life span: insights from a multimodal assessment study. *Psychol. Aging* **21:** 303.
86. Phillips, L.H., J.D. Henry, J.A. Hosie & A.B. Milne. 2008. Effective regulation of the experience and expression of negative affect in old age. *J. Gerontol. B Psychol. Sci. Soc. Sci.* **63:** P138.
87. Emery, L. & T.M. Hess. 2011. Cognitive consequences of expressive regulation in older adults. *Psychol. Aging* **26:** 388.
88. Opitz, P.C., L.C. Rauch, D.P. Terry & H.L. Urry. 2012. Prefrontal mediation of age differences in cognitive reappraisal. *Neurobiol. Aging* **33:** 645.
89. Shiota, M.N. & R.W. Levenson. 2009. Effects of aging on experimentally instructed detached reappraisal, positive reappraisal, and emotional behavior suppression. *Psychol. Aging* **24:** 890.
90. Winecoff, A. *et al.* 2011. Cognitive and neural contributors to emotion regulation in aging. *Soc. Cogn. Affect. Neurosci.* **6:** 165.
91. Scheibe, S. & F. Blanchard-Fields. 2009. Effects of regulating emotions on cognitive performance: what is costly for young adults is not so costly for older adults. *Psychol. Aging* **24:** 217.
92. Consedine, N.S. 2011. Capacities, targets, and tactics: lifespan emotion regulation from the perspective of developmental functionalism. In *Emotion Regulation and Well-Being*. I. Nyklíček, A. Vingerhoets & M. Zeelenberg, Eds.: 13–29. Springer. New York.
93. Critchley, H.D. *et al.* 2004. Neural systems supporting interoceptive awareness. *Nat. Neurosci.* **7:** 189.
94. Damasio, A. 2003. *Self: From Soul to Brain.* Vol. 1001. J. Ledoux, J. Debiec & H. Moss, Eds.: 253. New York Academy of Sciences. New York.
95. Lasch, H., D.O. Castell & J.A. Castell. 1997. Evidence for diminished visceral pain with aging: studies using graded

intraesophageal balloon distension. *Am. J. Physiol. Gastrointest. Liver Physiol.* **272:** G1.
96. Lagier, E. *et al.* 1999. Influence of age on rectal tone and sensitivity to distension in healthy subjects. *Neurogastroenterol. Motil.* **11:** 101.
97. Rayner, C.K., C.G. MacIntosh, I.M. Chapman & M. Horowitz. 2000. Effects of age on proximal gastric motor and sensory function. *Scand. J. Gastroenterol.* **35:** 1041.
98. Khalsa, S.S., D. Rudrauf & D. Tranel. 2009. Interoceptive awareness declines with age. *Psychophysiology* **46:** 1.
99. Good, C.D. *et al.* 2001. A voxel-based morphometric study of ageing in 465 normal adult human brains. *NeuroImage* **14:** 21.
100. Kalisch, T. *et al.* 2009. Impaired tactile acuity in old age is accompanied by enlarged hand representations in somatosensory cortex. *Cereb. Cortex* **19:** 1530.
101. Gibson, S.J. & F. Farrell. 2004. A review of age differences in the neurophysiology of nociception and the perceptual experience of pain. *Clin. J. Pain* **20:** 227.
102. Moore, A.R. & D. Clinch. 2004. Underlying mechanisms of impaired visceral pain perception in older people. *J. Am. Geriatr. Soc.* **52:** 132.
103. Cacioppo, J.T. *et al.* 2011. Could an aging brain contribute to subjective well-being? The value added by a social neuroscience perspective. In *Social Neuroscience: Toward Understanding the Underpinnings of the Social Mind.* A. Todorov, S.T. Fiske & D. Prentice, Eds.: 249. Oxford University Press. New York.
104. Mather, M. *et al.* 2004. Amygdala responses to emotionally valenced stimuli in older and younger adults. *Psychol. Sci.* **15:** 259.
105. Ochsner, K.N. *et al.* 2004. For better or for worse: neural systems supporting the cognitive down- and up-regulation of negative emotion. *NeuroImage* **23:** 483.
106. Liu, Y.W. *et al.* 2010. Analysis of regional MRI volumes and thicknesses as predictors of conversion from mild cognitive impairment to Alzheimer's disease. *Neurobiol. Aging* **31:** 1375.
107. Wright, C.I. *et al.* 2007. A functional magnetic resonance imaging study of amygdala responses to human faces in aging and mild Alzheimer's disease. *Biol. Psychiatry* **62:** 1388.
108. Wright, C.I. *et al.* 2006. Novel fearful faces activate the amygdala in healthy young and elderly adults. *Neurobiol. Aging* **27:** 361.
109. Leclerc, C.M. & E.A. Kensinger. 2008. Effects of age on detection of emotional information. *Psychol. Aging* **23:** 209.
110. Mather, M. & Knight, M.R. 2006. Angry faces get noticed quickly: threat detection is not impaired among older adults. *J. Gerontol. B Psychol. Sci. Soc. Sci.* **61:** P54.
111. Hahn, S., C. Carlson, S. Singer & S.D. Gronlund. 2006. Aging and visual search: Automatic and controlled attentional bias to threat faces. *Acta Psychol. (Amst.)* **123:** 312.
112. Hasin, D.S., R.D. Goodwin, F.S. Stinson & B.F. Grant. 2005. Epidemiology of major depressive disorder–results from the National Epidemiologic Survey on Alcoholism and Related Conditions. *Arch. Gen. Psychiatry* **62:** 1097.
113. Fiske, A., J.L. Wetherell & M. Gatz. 2009. Depression in older adults. *Annual Review of Clinical Psychology* **5:** 363.
114. Kendler, K.S., M. Gatz, C.O. Gardner & N.L. Pedersen. 2005. Age at onset and familial risk for major depression in a Swedish national twin sample. *Psychol. Med.* **35:** 1573.
115. Alexopoulos, G.S. 2006. The vascular depression hypothesis: 10 years later. *Biol. Psychiatry* **60:** 1304.
116. Alexopoulos, G.S. *et al.* 1997. 'Vascular depression' hypothesis. *Arch. Gen. Psychiatry* **54:** 915.
117. Krishnan, K.R.R., J.C. Hays & D.G. Blazer. 1997. MRI-defined vascular depression. *Am. J. Psychiatry* **154:** 497.
118. Sneed, J.R. & M.E. Culang-Reinlieb. 2011. The vascular depression hypothesis: an update. *Am. J. Geriatr. Psychiatry* **19:** 99.
119. Gunning-Dixon, F.M. *et al.* 2010. MRI signal hyperintensities and treatment remission of geriatric depression. *J. Affect. Disord.* **126:** 395.
120. Alexopoulos, G.S. *et al.* 2008. Microstructural white matter abnormalities and remission of geriatric depression. *Am. J. Psychiatry* **165:** 238.
121. Morimoto, S.S. *et al.* 2011. Executive function and short-term remission of geriatric depression: the role of semantic strategy. *Am. J. Geriatr. Psychiatry* **19:** 115.
122. Jorge, R.E., D.J. Moser, L. Acion & R.G. Robinson. 2008. Treatment of vascular depression using repetitive transcranial magnetic stimulation. *Arch. Gen. Psychiatry* **65:** 268.
123. Sneed, J.R. *et al.* 2008. The vascular depression subtype: evidence of internal validity. *Biol. Psychiatry* **64:** 491.
124. Herrmann, L.L., M. Le Masurier & K.P. Ebmeier. 2008. White matter hyperintensities in late life depression: a systematic review. *J. Neurol. Neurosurg. Psychiatry* **79:** 619.
125. Clark, L., S.R. Chamberlain & B.J. Sahakian. 2009. Neurocognitive mechanisms in depression: implications for treatment. *Annu. Rev. Neurosci.* **32:** 57.
126. Bush, G., P. Luu & M.I. Posner. 2000. Cognitive and emotional influences in anterior cingulate cortex. *Trends Cogn. Sci.* **4:** 215.
127. Gunning, F.M. *et al.* 2009. Anterior cingulate cortical volumes and treatment remission of geriatric depression. *Int. J. Geriatr. Psychiatry* **24:** 829.

Ann. N.Y. Acad. Sci. ISSN 0077-8923

ANNALS OF THE NEW YORK ACADEMY OF SCIENCES
Issue: *The Year in Cognitive Neuroscience*

# Perceptual foundations of bilingual acquisition in infancy

Janet Werker
University of British Columbia, Vancouver, British Columbia, Canada

Address for correspondence: Janet Werker, University of British Columbia, Department of Psychology, 2136 West Mall, Vancouver, BC V6T 1Z4. jwerker@psych.ubc.ca

**Infants are prepared by biology to acquire language, but it is the native language(s) they must learn. Over the first weeks and months of life, infants learn about the sounds and sights of their native language, and use that perceptual knowledge to pull out words and bootstrap grammar. In this paper, I review research showing that infants growing up bilingual learn the properties of each of the their two languages simultaneously, while nonetheless keeping them apart. Thus, they use perceptual learning to break into the properties of each of the two native languages. While the fundamental process of language acquisition is the same whether one or two languages are being acquired, cognitive advantages accrue from the task of language separation, and processing costs accrue from the more minimal input received in each of the two languages. I conclude by suggesting that when there are sufficient cues to which language is being used, the cognitive advantages that accrue from language separation enable the bilingual infant to move forward in language acquisition even in the face of processing costs.**

**Keywords:** language acquisition; bilingualism; infancy; perceptual foundations; review

## Introduction

One of the most remarkable feats of human development is the acquisition of language. Children begin understanding individual words by 6 months of age, typically produce their first words before their first birthday, begin combining words into short sentences during the second year of life, and become highly proficient language users shortly thereafter. But these feats would not be possible without prior learning of the perceptual properties of the native language. There is strong evidence to show that by birth infants have perceptual biases that orient them to language[1] and that enable them to discriminate its particulate components, as well as the cognitive machinery to learn the rules and regularities of the language to which they are exposed.[2] Some learning of the properties of the native language is already in place by birth, presumably from prenatal or immediate postnatal exposure.[3,4] Over the next several weeks and months, infants become more and more adept at attending to and optimally processing the properties of the native language, and become less sensitive to variation that is not meaningful in the native language. The ways in which changing perceptual sensitivities and learning machinery operate in tandem to launch acquisition of the native language is an increasingly important area of research. Only with perceptual knowledge of the rhythmical properties of the native language, of the speech sound categories that distinguish one possible word from another, and of the sequences of sounds that are allowable within a word and/or the statistical learning of other cues to segmentation, is a child able to pull out individual words and grammatical structures and map these on to meaning. Research over the past four decades has increasingly taught us just how this achievement unfolds, and how it sets the stage for language acquisition.

But not all children grow up learning just a single language. Indeed, estimates suggest that more than half of the world's population is bilingual, with many of those individuals learning more than one language from birth.[5] The infant who is growing up in a bilingual environment must learn the perceptual properties and rules of two native languages, and ultimately do so without confusing them. Depending on the languages being learned, the child

doi: 10.1111/j.1749-6632.2012.06484.x
 © 2012 New York Academy of Sciences.

has to acquire, for example, the rhythmical properties of each of two languages, the phonetic categories of each language, the phonotactic rules, and the word order, in addition to having two lexical entries for each concept, and in some cases, different conceptualizations of the world.

In this review, I will focus on the perceptual foundations of bilingual language acquisition. As such, I will take the bilingual infant up to the initial stages of lexical acquisition, focusing almost exclusively on infants from birth through 2 years of age (for an excellent review of bilingual language acquisition in toddlers and preschool-aged children, see Genesee & Nicoladis;[6] for earlier reviews of bilingual infants, see Sebastián-Gallés[7] and Werker, Byers-Heinlein, and Fennell[8]).

Infants who acquire two native languages pass the milestones in acquisition at approximately the same ages as do children who are acquiring only a single language (see Box A).[9–12] This is the case whether the two languages are spoken languages, two signed languages, or include both a spoken and a signed language. Many researchers interpret these similarities in rate as indicative of a single, primarily maturationally determined driver toward language acquisition,[12] and argue that the fact that total vocabulary is equivalent, even if smaller in each language in the bilingual, as evidence for the robustness of a common language-acquisition mechanism, even when the input per language is smaller. Others, however, focus on the differences, as in a recent report showing a small but significantly later age of first two-word utterances in bilinguals.[13] They, in turn, argue that considerations of input are important.

**Box A**

There is a long-standing concern, and one felt by many bilingual families, that if exposed to two languages early in life, children will become confused and mix the languages up in one big language pot. However, although there is some language mixing, there is little evidence of language confusion. Indeed, from the time they begin to acquire their first words, bilingual infants show comprehension and production of translation equivalents (e.g., a Spanish child using "casa" and "house").[90,91] By as young as 2-1/2 years of age, children growing up bilingual are able to actively choose which language to use when speaking to others. They will use language A if appropriate, language B if appropriate, and even switch between the two in a rule-governed way that matches their interlocutors, that is, if in their home environment language switching among active bilinguals is the rule.[92] Moreover, bilingual children this young can repair communication breakdowns caused by nonmatching language.[93] What is interesting is that they only use this kind of language repair when a language mismatch has caused the communication breakdown. In other situations, they try other kinds of repairs.

Although the cognitive machinery for acquiring two languages is essentially the same as that used for acquiring a single language, within this universality, I suggest that differences do accrue from having two rather than only one language of input. Two types of differences will be highlighted. One stems from the fact that in bilingual homes, infants typically receive less exposure to each individual language than in a home where only a single language is spoken. As we will see in this review, this affects both the speed and efficiency of processing, even in infancy. The other difference is one that ensues from having to separate the languages for simultaneous acquisition of two languages, and from having to keep them separate and minimize interference even after both languages are established. The requirement for language separation arguably recruits extralinguistic cognitive resources along with those used in processing and learning the language(s).

It is well documented in the adult literature that bilinguals show both processing costs and cognitive advantages. One processing cost that has been extensively documented is word retrieval: bilinguals perform more slowly and make more mistakes,[14] particularly in speeded tasks, than do monolinguals.[15] Moreover, this processing cost is seen even in the dominant language.[16]

Adult[17–19] and child[20–22] bilinguals show a number of cognitive advantages over monolinguals, particularly in executive functioning[a] (see Ref. 24 for a

[a]In the adult literature, the processing disadvantages are discussed in terms of selection between the two languages (e.g., Marian & Spivey;[23] see Kroll, Dussisas, Bogulski & Kroff[24] for a review) rather than the amount of input.

review). Recent studies have shown that already by 7 months of age, bilingual learning infants also show enhanced cognitive control compared to monolinguals. Whereas both monolingual and bilingual infants can learn to turn toward one type of sound for a visual reward, bilingual infants are better able than monolingual infants to switch and learn to subsequently respond to a different sound.[25,26] By 24 months, bilingual toddlers show advantages in the Stroop task,[27] a standard test of executive function.

Below, I review in detail recent work with infants on the perceptual foundations of bilingual language acquisition. I begin with a consideration of the cues used for language separation, how such cues might bootstrap acquisition, how and when native speech sound categories are established and how this directs word learning, and end with a consideration of whether perception has influenced the cognitive biases that guide word learning. In every case, I first review what we know about monolingual acquisition, and use this as a lens to present work with bilinguals. Although both monolingual and bilingual acquisition are likely equally common and equally natural, comparing one with the other allows a consideration of the possible costs for the bilingual infant of a decreased amount of input in each language, along with the potential cognitive advantages of having to keep the two languages distinct.

## Language discrimination and preference

The languages of the world differ on many different properties, with one of the most salient being rhythm. Languages such as English, German, and Czech are described as stress timed, a rhythmic pattern that entails having both strong and weak syllables (e.g., LANguage or SPEcies), relative isochrony from one strong syllable to the next in running speech, vowel reduction in nonstressed syllables, and fairly complex syllable structure, including consonant clusters.[28] Syllable-timed languages such as French, Spanish, and Italian have less syllable-level stress, less complex syllable forms, and relative isochrony from one syllable to the next in running speech (see Ramus, Nespor, and Mehler for one quantification[29]). These rhythmical properties are highly correlated with sentence-level word order,[30] and processing of them has been hypothesized to be the infants' first entry into the grammar of the native language structure. At birth, monolingual neonates can discriminate languages from different rhythmical classes,[31] even when the two languages are unfamiliar. Thus, the ability to discriminate languages based on rhythm seems to be present independent of prenatal listening experience. However, at birth monolingual infants nonetheless show a preference for listening to their native language, thus evidencing an effect of experience.[3]

The ability to keep the two native languages separate, even when they are from the same rhythmical class, emerges surprisingly early in development. Bosch and Sebastián-Gallés[32] were the first to find evidence of language discrimination in young bilingual infants by showing that Spanish-Catalan infants aged 4 months could detect a change, in a habituation task, from one of their syllable-timed native languages to the other. Moreover, although language separation was easy for these infants, they showed equal interest in listening to each of their native languages.

In some situations, the mother may speak only one language, whereas the father or other members of the family speak a different language. But in many cases, the mother has been speaking two languages throughout pregnancy. This again raises the question of how the experience of hearing two languages throughout gestation affects language discrimination and preference. Such experience could, on the one hand, enhance language separation. But on the other hand, it could also theoretically interfere with such separation, and establish a big category of all familiar languages.

To examine this question, we tested rhythmical language discrimination in neonates who had been exposed to two rhythmically distinct languages in utero. We presented both filtered English speech (stress timed) and filtered Tagalog (Filipino) speech (syllable timed) to both infants who had heard only English in utero and to bilingual English-Tagalog neonates whose mothers had spoken both languages approximately equally throughout their pregnancy. We habituated the neonates to one language in a contingent sucking procedure, and then presented them with the other languages. The results were unequivocal: both the English monolingual and the English-Tagalog bilingual neonates robustly discriminated the two languages. In a subsequent test of preference, we found that while the monolingual English-exposed neonates preferred listening to English, the bilingual infants listened to both languages

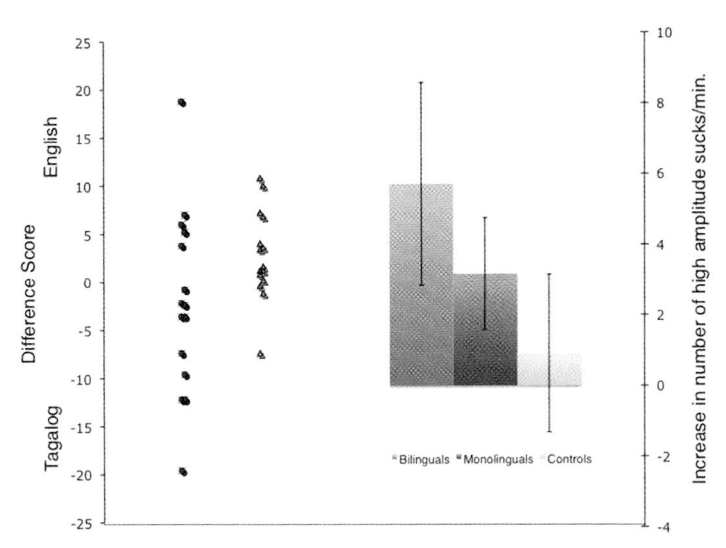


**Figure 1.** In the high-amplitude sucking (HAS) procedure, an infant's average sucking strength to a pacifier is measured in the first baseline minute. In subsequent minutes, every strong suck is followed by the presentation of a sound. In the preference task (left), in alternating minutes (5 of each), infants had the chance to listen to Tagalog or English. The number of HA sucks per minute was recorded. The English monolingual infants chose to listen more to English (green triangles), but the bilingual English-Tagalog infants listened equally to both (red circles). In the discrimination task (right), neonates were presented with filtered speech from one language (either English or Tagalog) for several minutes until they habituated (the number of HA sucks for 2 consecutive minutes dropped to 50% of the number during the highest 2 minutes). They were then presented with filtered speech from the other language (experimental group) or the same language (control group). Both the bilingual English-Tagalog and monolingual English infants discriminated the change in language as shown by the recovery in number of sucks in the 2 posthabituation minutes, whereas neonates in the control group did not. This figure has been modified from Figure 2 in Byers-Heinlein & Werker.[33]

equally. Hence, although listening experience can shape listening preferences, it cannot overwrite the sensitivity to the rhythmical cues that distinguish the languages. At a practical level, this allows the bilingual infant to pay close attention to both of her native languages while also keeping them apart. At a theoretical level, the continuing experience-independent sensitivity to rhythmical differences leaves the bilingual neonate with the tools required for later using rhythm to bootstrap acquisition of each of her native languages (Figs. 1 and 2).

Recently, we found that bilingual infants can use their sensitivity to the rhythm of language to help figure out its basic word order.[94] The languages of the world differ in their basic word order, with the two most canonical forms being subject–verb–object (SVO), as in languages such as French or English (e.g., "The boy threw my ball"), and subject–object–verb (SOV), as in languages such as Japanese or Turkish. Basic word order is highly correlated with the order of other constituents in the language. SVO languages, for example, tend to have articles (the, an) or pronouns (his, her) in front of nouns and tend to have prepositions (with, from) in noun phrases. SOV languages tend to have the article or pronoun occur after the noun and tend to have postpositions. This results in two kinds of statistics that are perceptually available to prelinguistic infants. First, in SVO languages, frequent words occur before infrequent words, whereas SOV languages have the opposite word order. By 7 months, infants can use frequency to parse an artificial language into phrases.[2,34] Second, prosody is correlated with word order. In VO languages, phrases tend to be trochaic (an unstressed syllable followed by a stressed syllable) and prosodic prominence is indicated by duration.[35] Thus, there is a short-long rhythm to noun phrases. In OV languages, prosodic prominence in a phrase is indicated by pitch, resulting in a high-low grouping in noun phrases. In a recent study, we found that at 7 months, bilingual infants[b] who are growing up in languages with two different word orders (so frequency is not a

[b]The infants were bilingual with English, an SVO language, and an SOV language (Japanese, Korean, Hind/Punjabi, Farsi, or Turkish).

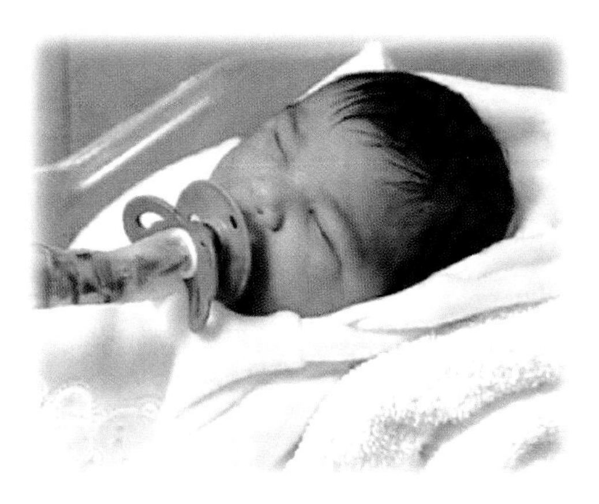

**Figure 2.** The photo shows a newborn sucking on the pacifier that is connected to a pressure transducer for measurement and stimulus control in the HAS procedure.

predictable cue) are able to use prosody to segment noun phrases (J. Gervain & J.F. Werker, unpublished results). Hence, learning the rhythmic properties of each of their two languages seems to prepare bilingual infants to acquire the word order of each of their two languages. Moreover, they are sensitive to a cue—prosody—that monolingual infants seem not to use.

## Language discrimination using visual information

Of interest, heard language is not the only cue bilingual babies use to keep their two languages apart. They also use the cues seen in talking faces. When we speak, the shape of our mouths and the related muscles involved reflect the sounds we are producing,[36] and babies are sensitive to the match between heard and seen speech sounds.[37,38] Moreover, the timing of the opening and closing of the jaw tracks the rhythmicity of the language.[36] We asked whether infants could use their interest in the movement of talking faces to discriminate one language from another. We recorded three bilingual women reciting sentences from a children's story. We then presented these to monolingual infants aged 4, 6, and 8 months in a discrimination paradigm with the sound turned *off*. In each trial, the infant saw one face producing one sentence. Each of the three faces was presented in turn, always producing a new sentence, in subsequent trials until the infant habituated. We then showed the control group of infants the same faces reciting new sentences (as before) in the same language, and showed the experimental group the same faces reciting new sentences, but in the opposite language.

The results were startling. At 4 and 6 months, the monolingual English infants were able to discriminate the change, whereas the infants in the experimental but not control groups showed an increase in looking following habituation. However, at 8 months the monolingual infants failed, no longer paying attention to this cue. We then tested French–English bilingual-learning infants at 6 and 8 months and found that the ability to discriminate their two native languages by watching silent, talking faces was maintained. One can speculate that such sensitivity could be very useful to the bilingual child. Thus, an initial, universal perceptual bias to use visual cues to discriminate languages was attenuated in infants who experienced only a single language of input, and maintained in infants who experienced both, again shows the contribution of both universal beginnings and input-driven processes in preparing the bilingual infants for dual language acquisition.

## Cognitive advantages

One question raised from the visual language discrimination work is whether the bilingual French–English infants maintained the ability to discriminate the two languages visually at 8 months because they were familiar with the properties in talking faces that correspond to French versus English, or whether it was because bilingual-learning infants simply pay attention to possible cues in their world that might help them distinguish one language from another. To address this question, we tested another group of bilingual infants at 8 months, Spanish-Catalan–learning infants, on their ability to discriminate visual English from visual French, and compared their performance with monolingual Catalan and/or Spanish infants. As expected, the monolingual infants were unable to discriminate the two unfamiliar languages at 8 months of age. However, the bilingual Spanish–Catalan infants succeeded—even though neither language was familiar.[39] These results suggest that bilingual experience may result in heightened perceptual vigilance, at least in the language domain. Perhaps this initial heightened perceptual attentiveness is an important factor contributing to the emergence of the broader cognitive advantages seen in bilingual infants and adults.

## Language context as an anchor

To the extent that rhythmical cues are present, or there is information from talking faces or some other source, it might be increasingly easier for the bilingual-learning infant to separately track and learn about the properties of each of her two languages. Although there is little evidence that adult bilinguals use such cues to help separate each of their native languages,[40] recent work using artificial-language–learning manipulations has shown that in the presence of two voices,[41] or two difference faces,[42] adults can learn to segment nonsense syllables in two different ways. If distinct cues are effective in learning, bilingual learning infants might be able to rely on what they have already learned about each language (e.g., its rhythmical properties or its phonotactics) as a context for separately tracking and learning more properties of each of their native languages.[43] I will return to this theme periodically in this paper.

## Establishing and using native phoneme categories

### *Monolingual infants*

The fundamental units that distinguish one word from another, as in the words "bog" versus "dog" versus "fog," or "bog" versus "bag," are called *phonemes*. The languages of the world differ in the number and precise phoneme categories that they use. In English, for example, there are six stop consonants as exemplified in the syllables /ba/, /da/, /ga/, /pa/, /ta/, and /ka/, whereas in Hindi there are 16. Similarly, English makes a distinction between /ra/ and /la/, whereas Japanese has only a single phoneme, which is intermediate between the two. Infants begin life discriminating many of the sounds of the world's languages. Development during the first year of life includes sharpening those sound categories that are used in the native language,[44–46] and "narrowing" perception such that those distinctions not used in the native language cease to be readily discriminated.[47–49]

One focus of current research is to determine what the learning mechanisms are that enable infants to establish their native phoneme repertoire. In my lab, we have examined two broad classes of mechanisms. One involves relatively passive tuning to the most frequently experienced input, for example, by distributional learning.[50] A second that we call "acquired distinctiveness"[51,52] could be a closer route to meaning: the cooccurrence of two phones with two different objects could help pull them apart, whereas the cooccurrence of two phones with a single object could help collapse the distinction.[53]

The speech sound categories established in the first year of life ultimately guide word learning. Thus, by 18 months a child growing up in Dutch, for example, will treat the difference between a long /aa/ and a short /a/ (a distinction used in Dutch) as referring to two different words, whereas a child growing up in English has learned that this distinction is not useful and will treat both pronunciations as referring to the same object.[54]

### *Bilingual infants*

Infants who grow up bilingual need to be able to discriminate the speech sound contrasts used in each of their languages, while also distinguishing a particular sound in one language from that in the other. Research shows that they do this remarkably well: by the end of the first year of life, they show robust discrimination of the speech-sound distinctions in each of their two native languages.[55–57] Moreover, by this age infants also discriminate the same phone me as used in each of their languages. For example, at 10–12 months, French–English bilingual infants can discriminate a /d/ pronounced with a French accent from one pronounced with an English accent.[58]

## Impact on speed and efficiency of processing

Although bilingual infants do establish two sets of phonetic categories, the timing of a robust change from "universal" to "native listening" might be different. In the first experimental paper testing bilingual phonetic discrimination, Bosch and Sebastián-Gallés[57] compared Spanish and Catalan monolingual-learning infants to Spanish–Catalan bilingual infants on their ability to discriminate the /e/-/E/ (as in the English words "late" versus "let") distinction that is used in Catalan but not in Spanish.[57] As expected, at 4 months of age, infants in all three language groups discriminated the distinction, and by 8 months of age, while the Catalan monolingual infants maintained discrimination of their native distinction, the Spanish monolingual infants were no longer successful. Surprisingly, however, the Spanish–Catalan bilingual infants also failed at

8 months, even though they were hearing the distinction in one of their native languages. They succeeded again at 12 months, leading to the suggestion that there may be a temporary delay in bilinguals as they traverse the path of establishing two sets of native phonetic distinctions while also keeping them apart. Similar results of a period of potential confusion in bilingual infants have been reported for other phonetic contrasts,[59,60] and even with older bilingual infants.[61]

The first explanation for these results was based on the notion of distributional learning. There is an /e/-/E/ distinction in Catalan but only an /e/ vowel in Spanish (close to, but not exactly like the Catalan one, and overlapping somewhat with the Catalan /E/). Hence, a Spanish–Catalan bilingual might hear so many more /e/ than /E/ vowels, and with some of the /e/ vowels overlapping the /E/ category, the distinction between /e/ and /E/ could be temporarily swamped.[60] Alternatively, there are many cognates in the two languages (similar sounding words that mean the same thing, as in the Spanish "pera" and the Catalan "pEra," both of which mean "pear"—also a cognate in English). Thus, if acquired distinctiveness plays a role in learning native phonetic categories,[53] the presence of cognates referring to the same object in each language could provide conflicting information to the bilingual learner.[60,62] (For a more in-depth review and a slightly different argument, see Ref. 63.) Relatively, it has been shown that bilingual infants whose two languages are from different rhythmical classes are able to discriminate the /e/-/E/ distinction at 8 months, even when tested in a procedure similar to the one in which same-aged Spanish–Catalan bilingual infants failed. This is taken as evidence that bilingual infants rely on higher-level cues, in this case rhythm, to separately track the phonetic characteristics of each language.[64]

Context also plays a role in bilingual infants' use of native-language phonemic categories to guide-word-learning and word recognition. In a paper published in 2007, we reported a later age of success in bilingual than in monolingual infants. When tested in an associative-word-learning task in which they were shown one object repeatedly paired with the nonsense word "bih" and another object paired with the nonsense word "dih," and then shown two test trials—one with the "same" pairing (word A, object A), and one with the pairing "switched" (familiar word, familiar object, but now word A, object B), the monolinguals looked longer to the switch than in the same trial at 17 months, but the bilingual infants did not do so until 20 months.[65] However, a subsequent study[66] reported that if the pronunciation of the individual words is appropriate to signal a monolingual English, a monolingual French, or a bilingual English–French context, infants in each of those groups succeed at the same, 17-month age. It is only when the pronunciation does not match, e.g., English pronunciation tested with English–French bilinguals, that infants show difficulty. More recently, Fennell and Byers-Heinlein[67] showed that when first given sentences that specify the language being used, bilingual infants succeed at learning minimally different words even if the pronunciation of the individual words is slightly accented.

Although the above findings show that bilingual infants rely on context to let them know which set of phonological categories to use, it is still unclear whether bilinguals have as fully developed representations of the phonological categories of each language and simply wait for contextual cues as to which language to activate, or whether the representations are less robust at the same age, awaiting accrual of more input to become more firmly established. In electrophysiological recordings of the mismatched negativity (MMN), a negative event-related potential recorded around 250 ms after a change from a repeated standard to an oddball to test phonetic discrimination,[68] clear differences are seen between monolingual and bilingual infants. Whereas the MMN to a phonetic change shows increasing maturity between 3 and 36 months in monolingual infants—gradually changing from more positive to more negative and from posterior to frontal—the directionality and topography of the MMN response changes in less consistent ways during this time period in bilingual infants.[69] Similar findings were reported by Garcia-Sierra *et al.*[70] in a study comparing monolingual English infants to bilingual English–Spanish infants on their ability to discriminate the English and Spanish /da/-/ta/ distinctions. Overall, the MMN response matured more slowly in the bilingual than in the English infants. Indeed, while an ERP indicative of discrimination was seen in the monolingual infants at 6–9 months, nothing was evident in the bilingual infants until 10–12 months, suggesting a representational difference in the native phonetic categories in the two groups at the younger age (also see Petitto *et al.*[71] for complementary results using optical imaging).

In the Garcia-Sierra *et al.* study,[70] the amount of input in each language proved to be very important. The maturity of the MMN response was more mature for Spanish if the bilingual infants heard more Spanish than English, and visa versa. Moreover, vocabulary size in each individual language was correlated with MMN maturity. These results would suggest that while bilingual infants may be able to use context to help them disambiguate a difficult task, the amount of input in each language also independently influences the elaboration of the representation.

Results suggesting a difference in the robustness of the representation are also found in the word-recognition literature. Whereas there is a greater left than right hemispheric response over frontal and temporal cortex to known words in monolingual infants aged 19–22 months, it is less lateralized, and slower, in bilingual infants.[72] Moreover, the degree of difference is predicted by vocabulary size in the nondominant language. Behaviorally, monolingual English toddlers (30 months) are faster than bilingual Spanish–English infants at orienting away from the incorrect and toward the correct referent when presented with a single spoken label, and shown a picture of both a matching and a nonmatching object.[73] This difference is best predicted by the amount of input the infants receive in each language, and by their current vocabulary in each language as measured on the MacArthur-Bates Communicative Development Inventory (CDI) parental vocabulary checklist. Similarly, bilingual Spanish–Catalan toddlers aged 14–24 months are more likely than monolingual Catalan toddlers to treat a mispronunciation (from a Catalan /e/ to a Catalan /E/) of a known Catalan word as acceptable and still look at the "matching" object. This effect is also moderated by the amount of input in each language such that those bilingual infants who receive relatively more Catalan input are more likely to perform like their monolingual Catalan peers.[74]

The somewhat contradictory results reviewed previously make more sense when consideration is given to the possibility that two independent factors may be contributing. There are likely consequences in processing efficiency that result from less input from each individual language. At the same time, however, there may be cognitive advantages that allow the bilingual-learning infant to use contextualized perceptual cues to facilitate performance. As such, the bilingual infant is most likely to show processing costs when in an environment in which there are no contextual cues indicating which language should be used.

## Word-learning biases

A well-studied issue in language acquisition is the problem of induction. When an infant hears a word like "cat," how does he or she know that it refers to the whole cat rather than a part of the cat, such as an ear, a property of the cat such as its color, or an action that the cat may be involved in? A number of word-learning constraints have been proposed to help explain why it is that in the initial stages of acquisition infants seem to treat words as nouns labeling whole objects, and more specifically as labels for categories, and only after such referents are established do they consider alternative meanings.[75]

One constraint that has been studied is commonly referred to as "mutual exclusivity." This is the assumption that each object category has only one label.[76] Hence, when first encountering a cat, the child will assume that the label "cat" refers to the entire category, but if they later encounter the same cat along with another animal, for example, a "dog," and hear the label "dog," they will reject that label as a second label for the cat, and treat it instead as a label for the other object. Mutual exclusivity has been shown to be evident by 17 months of age.[77] While there is agreement on the existence of this word-learning bias in monolingual infants, there is no agreement on what it means or where it comes from developmentally.[78] In this case, research with multilingual infants can help address a fundamental question in acquisition. While mutual exclusivity can be seen to be adaptive for monolingual-learning infants, bilingual infants regularly encounter—and need to learn—more than one label for the same object, as they receive labels in each of their languages. Moreover, bilingual children often hear language with a lot of "code switching," that is, words from the other language inserted in a rule-governed and regular way, but nonetheless resulting in mixed input. Research with bilingual children from the preschool to school-aged years indicate that by the time they reach school age, both monolingual and bilingual children understand that two objects can have the same label if the labels come from different languages,[79] but while still preschoolers, bilinguals can sometimes become confused and reject a second label even if it is in their second language.[80]

Two recent studies have tested mutual exclusivity in bilingual in comparison with monolingual infants.[81,82] In both of these studies, a disambiguation paradigm (as in Ref. 77) was used to test mutual exclusivity. In this paradigm, infants are tested in a two-choice preference task that includes two kinds of trials: known–known and known–unknown. In the first type, two known objects are presented side by side, and a label that corresponds to one is presented. Here, the child is expected to look longer to the matching object. In the second type, the child sees a known and an unknown object, and in disambiguation trials, hears a novel label. Mutual exclusivity is assumed if the child looks longer to the novel object in the presence of the novel label.

Mutual exclusivity was not as pronounced in bilingual as in monolingual infants in the Byers-Heinlein and Werker study,[81] and not evident at all in trilingual-learning infants. In the Houston-Price study,[82] mutual exclusivity was not seen in the bilinguals. In the Byers-Heinlein and Werker sample, no correlation with degree of mutual exclusivity was found between either the amount of input and/or the vocabulary size in each language. Hence, we claimed that the driving force in whether mutual exclusivity would be seen or not was the overall structure of the lexicon across the languages. If children had experienced more than one label for the objects in their world, they were more likely to entertain the possibility, even in an experimental task, that the same object category could be given more than a single label. In the Houston-Price *et al.* study, there was an effect of vocabulary size, but only in the monolingual sample. Those infants with a larger English vocabulary were more likely to show mutual exclusivity. Based on these findings, both groups of researchers have argued that rather than being a built-in constraint that guides word learning, mutual exclusivity may be a bias that emerges across development in monolingual infants as a function of establishing a lexicon made up of one-to-one word–concept mappings.

## Summary

The perceptual systems provide the first entry point into acquisition of the native language. In this review, I have presented evidence showing that from birth, infants growing up with two languages can use perceptual cues to begin to separate the languages and to learn the sound properties of each. This in turn prepares them for word learning, and even for the first steps in bootstrapping grammar.

While the milestones achieved in bilingual acquisition are largely parallel to those achieved in monolingual acquisition, there is evidence of both processing challenges and cognitive advantages in bilingual acquisition. The challenges seem to stem, at least in part, from the fact that with two native languages, there is less input in each. This results in less well-established representations, processing inefficiencies, and in some cases, slight delays in acquisition. The cognitive advantages seem to come from having to separate, and keep separate, the two native languages while acquiring the properties of each. The cognitive advantages discussed included better attention to perceptual details that might distinguish a talking face, heard speech, or even individual sounds from one language versus another, as well as the ability to better use context to determine which language is being used. Thus, while the fundamental mechanisms supporting language acquisition are the same in the bilingual and in the monolingual infant, the input does play a role. This role can be seen in the brain systems involved in language processing and use, in speed of processing, in total vocabulary size in each language, in word learning and recognition, and perhaps more fundamentally in the biases children use to figure out the meaning of words.

## Future directions

To date, the majority of studies with bilingual-learning infants have involved infants acquiring two spoken languages. There is, however, increasing work on bimodal bilinguals—who have one spoken and one signed language.[83] Initial steps in acquisition seem to be identical in bimodal/bilingual–learning infants to the steps in acquisition of a spoken language,[84] and involve the same language areas in the brain.[12] Although there is still significant interference between even spoken and signed languages in the adult brain, there may be less interference from the second language in signing-speaking adults,[85] in part because they are governed by different modalities.[86] Thus, the processing costs may be diminished. If the conclusions drawn in this review are correct, the pattern of results with adult bimodal/bilinguals would predict that the role of the input would play out differently in bimodal/bilingual infants. They may not have the processing challenges that the oral bilingual child has, and in turn they also may not have as pronounced cognitive advantages.

As evidenced in this review, there is increasing evidence that the amount of input in each language is important for fully understanding the first steps in bilingual acquisition. In addition to the several excellent parent-report questionnaires that are being used to estimate the amount of input in each language,[87] new tools are being developed. One is a language-mixing questionnaire developed by Byers-Heinlein[88] that provides a highly reliable estimate of the degree to which the bilingual input comes from two clearly distinct sources, or is more likely to come via mixing within individual speakers. Data from this questionnaire are already helping to advance theoretical understanding and empirical results.[13] A second is an exciting tool, LENA, which records all speech heard by an individual child over the period of time the recording device is worn. New research using this tool reveals much more nuanced information about how the qualitative properties of input speech, in addition to simple quantity, influence language development.[89] Recording tools such as LENA also have promise for better characterization of the language input in bilinguals. With betterinfluences characterization of the input, we will be better situated to address increasingly precise theoretical predictions.

## Conflicts of interest

The author declares no conflicts of interest.

## References


1. Vouloumanos, A. & J.F. Werker. 2007. Listening to language at birth: evidence for a bias for speech in neonates. *Dev. Sci.* **10:** 159–164.
2. Gervain, J., F. Macagno, S. Cogoi *et al.* 2008. The neonate brain detects speech structure. *Proc. Natl. Acad. Sci. USA* **105:** 14222–14227.
3. Moon, C., R.P. Cooper & W.P. Fifer. 1993. Two-day-olds prefer their native language. *Infant Behav. Dev.* **16:** 495–500.
4. May, L., K. Byers-Heinlein, J. Gervain, *et al.* 2011. Language and the newborn brain: does prenatal language experience shape the neonate neural response to speech? *Front. Psychol.* **2:** 222.
5. Grosjean, F. 2008. *Studying Bilinguals.* Oxford University Press. New York.
6. Genesee, F. & E. Nicoladis. 2008. Bilingual first language acquisition. In *Blackwell Handbook of Language Development.* E. Hoff & M. Shatz, Eds.: 324–342. Blackwell Publishing Ltd. Oxford, UK.
7. Sebastián-Gallés, N. 2010. Bilingual language acquisition: where does the difference lie? *Hum. Dev.* **53:** 245–255.
8. Werker, J.F., K. Byers-Heinlein & C.T. Fennell. 2009. Bilingual beginnings to learning words. *Philos. T. Roy. Soc. B* **364:** 3649–3663.
9. Holowka, S., F. Brosseau-Lapré & L.A. Petitto. 2002. Semantic and conceptual knowledge underlying bilingual babies' first signs and words. *Lang. Learn.* **52:** 205–262.
10. Paradis, J. & F. Genesee. 1996. Syntactic acquisition in bilingual children. *Stud. Second Lang. Acquis.* **18:** 1–25.
11. Pearson, B.Z. 2009. Children with two languages. In *Handbook of Child Language.* E. Bavin, Ed.: 379–398. Cambridge University Press, Cambridge.
12. Petitto, L.A. & I. Kovelman. 2003. The bilingual paradox: how signing-speaking bilingual children help us to resolve it and teach us about the brain's mechanisms underlying all language acquisition. *Learn. Lang.* **8:** 5–18.
13. Hoff E., C. Core, S. Place, *et al.* 2012. Dual language exposure and early bilingual development. *J. Child Lang.* **39:** 1–27.
14. Gollan, T.H., R.I. Montoya, C. Fennema-Notestine, *et al.* 2005. Bilingualism affects picture naming but not picture classification. *Mem. Cogn.* **33:** 1220–1234.
15. Bialystok, E., F.I.M. Craik, D.W. Green, *et al.* 2009. Bilingual minds. *Psychol. Sci. Pub. Interest* **10:** 89–129.
16. Ivanova, I. & A. Costa. 2008. Does bilingualism hamper lexical access in speech production? *Acta Psychol.* **127:** 277–288.
17. Bialystok, E. 2010. Bilingualism. *Wiley Interdisciplinary Reviews: Cogn. Sci.* **1:** 559–572.
18. Costa, A., M. Hernández & N. Sebastián-Gallés. 2008. Bilingualism aids conflict resolution: evidence from the ANT task. *Cognition* **106:** 59–86.
19. Hernández, M., A. Costa, L.J. Fuentes, *et al.* 2010. The impact of bilingualism on the executive control and orienting networks of attention. *Biling. Lang. Cogn.* **13:** 315–325.
20. Bialystok, E. 1999. Cognitive complexity and attentional control in the bilingual mind. *Child Dev.* **70:** 636–644.
21. Bialystok, E., R. Barac, A. Blaye, *et al.* 2010. Word mapping and executive functioning in young monolingual and bilingual children. *J. Cog. Dev.* **11:** 485–508.
22. Carlson, S.M. & A.N. Meltzoff. 2008. Bilingual experience and executive functioning in young children. *Dev. Sci.* **11:** 282–298.
23. Marian, V. & M. Spivey. 2003. Competing activation in bilingual language processing: within-and between-language competition. *Biling. Lang. Cogn.* **6:** 97–116.
24. Kroll, J.F., P.E. Dussias, C.A. Bogulski, *et al.* 2012. Juggling two languages in one mind: what bilinguals tell us about language processing and its consequences for cognition. *Psychol. Learn. Motiv.* **56:** 229–262.
25. Kovács, Á.M. & J. Mehler. 2009. Cognitive gains in 7-month-old bilingual infants. *P. Natl. Acad. Sci. USA* **106:** 6556.
26. Kovács, Á.M. & J. Mehler. 2009. Flexible learning of multiple speech structures in bilingual infants. *Science* **325:** 611–612.
27. Poulin-Dubois, D., A. Blaye, J. Coutya, *et al.* 2011. The effects of bilingualism on toddlers' executive functioning. *J. Exp. Child Psychol.* **108:** 567–579.

28. Abercrombie, D. 1967. *Elements of General Phonetics*. Edinburgh University Press. Edinburgh.
29. Ramus, F., M. Nespor & J. Mehler. 1999. Correlates of linguistic rhythm in the speech signal. *Cognition* **73:** 265–292.
30. Shukla, M. & M. Nespor. 2010. Rhythmic patterns cue word order. In *The sound patterns of syntax*. N. Erteschik-Shir & L. Rochman, Eds.: 174–189. Oxford University Press. Oxford.
31. Nazzi, T., J. Bertoncini & J. Mehler. 1998. Language discrimination by newborns: toward an understanding of the role of rhythm. *J. Exp. Psychol. Human*. **24:** 756.
32. Bosch, L. & N. Sebastián-Gallés. 2001. Evidence of early language discrimination abilities in infants from bilingual environments. *Infancy* **2:** 29–49.
33. Byers-Heinlein, K., T.C. Burns & J.F. Werker. 2010. The roots of bilingualism in newborns. *Psychol. Sci.* **21:** 343.
34. Bion, R.A., S. Benavides-Varela & M. Nespor. 2011. Acoustic markers of prominence influence infants' and adults' segmentation of speech sequences. *Lang. Speech.* **54:** 123–140.
35. Nespor, M. & I. Vogel. 2007. *Prosodic Phonology*. Walter de Gruyter Berlin.
36. Munhall, K.G. & E. Vatikiotis-Bateson. 1998. The moving face during speech communication. In *Hearing by Eye, Part 2: Advances in the Psychology of Speechreading and Auditory-Visual Speech*. R. Campbell, B. Dodd & D. Burnham, Eds.: 123–139. Taylor & Francis—Psychology Press. Sussex, UK.
37. Kuhl, P.K., & A.N. Meltzoff. 1982. The bimodal development of speech in infancy. *Science* **218:** 1138–1141.
38. Patterson, M.L. & J.F. Werker. 2003. Two-month-old infants match phonetic information in lips and voice. *Dev. Sci.* **6:** 191–196.
39. Sebastián-Gallés, N., W. Weikum, B. Albareda-Castellot, *et al.* A bilingual advantage in visual language discrimination in enfancy. *Psych. Sci.* In press.
40. Morford, J.P., E. Wilkinson, A. Villwock, *et al.* 2010. When deaf signers read English: do written words activate their sign translations? *Cognition* **118:** 292.
41. Weiss, D.J., C. Gerfen & A.D. Mitchel. 2009. Speech segmentation in a simulated bilingual environment: a challenge for statistical learning? *Lang. Learn. Dev.* **5:** 30–49.
42. Mitchel, A. & D.J. Weiss. 2010. What's in a face? Visual contributions to speech segmentation. *Lang. Cognitive Proc.* **25:** 456–482.
43. Curtin, S., K. Byers-Heinlein & J.F. Werker. 2011. Bilingual beginnings as a lens for theory development: PRIMIR in focus. *J. Phonetics* **39:** 492–504.
44. Kuhl, P.K., E. Stevens, A. Hayashi, *et al.* 2006. Infants who a facilitation effect for native language phonetic perception between 6 and 12 months. *Dev. Sci.* **9:** F13–F21.
45. Narayan, C.R., J.F. Werker & P.S. Beddor. 2010. The interaction between acoustic salience and language experience in developmental speech perception: evidence from nasal place discrimination. *Dev. Sci.* **13:** 407–420.
46. Sato, Y., Y. Sogabe & R. Mazuka. 2010. Discrimination of phonemic vowel length by Japanese infants. *Dev. Psych.* **46:** 106–19.
47. Best, C., G. McRoberts, R. LaFleur, *et al.* 1995. Divergent developmental patterns for infants' perception of two nonnative consonant contrasts. *Infant Behav. Dev.* **18:** 339–350.
48. Mattock, K. & D. Burnham. 2006. Chinese and English infants' tone perception: evidence for perceptual reorganization. *Infancy* **10:** 241–265.
49. Werker, J.F. & R.C. Tees. 1984. Cross-language speech perception: evidence for perceptual reorganization during the first year of life. *Infant Behav. Dev.* **25:** 49–63.
50. Maye, J., J.F. Werker & L.A. Gerken. 2002. Infant sensitivity to distributional information can affect phonetic discrimination. *Cognition* **82:** B101–B111.
51. Hall, G. 1991. *Perceptual and Associative Learning*. Clarendon Press/Oxford University Press. Oxford.
52. Lawrence, D.H. 1949. Acquired distinctiveness of cues: I. Transfer between discriminations on the basis of familiarity with the stimulus. *J. Exp. Psychol.* **39:** 770–784.
53. Yeung, H.H. & J.F. Werker. 2009. Learning words' sounds before learning how words sound: 9-month-olds use distinct objects as cues to categorize speech information. *Cognition* **113:** 234–243.
54. Dietrich, C., D. Swingley & J.F. Werker. 2007. Native language governs interpretation of salient speech sound differences at 18 months. *Proc. Natl. Acad. Sci. USA* **104:** 16027–16031.
55. Albareda-Castellot, B., F. Pons & N. Sebastián-Gallés. 2011. The acquisition of phonetic categories in bilingual infants: new data from an anticipatory eye movement paradigm. *Dev. Sci.* **14:** 395–401.
56. Burns, T.C., K.A. Yoshida, K. Hill, *et al.* 2007. The development of phonetic representation in bilingual and monolingual infants. *Appl. Psycholinguist.* **28:** 455.
57. Bosch, L. & N. Sebastián-Gallés. 2003. Simultaneous bilingualism and the perception of a language-specific vowel contrast in the first year of life. *Lang. Speech* **46:** 217–243.
58. Sundara, M., L. Polka & M. Molnar. 2008. Development of coronal stop perception: bilingual infants keep pace with their monolingual peers. *Cognition* **108:** 232–242.
59. Bosch, L. & N. Sebastián-Gallés. 2003. Language experience and the perception of a voicing contrast in fricatives: infant and adult data. In *International Congress of Phonetic Sciences*. Barcelona, 1987–1990.
60. Sebastián-Gallés, N. & L. Bosch. 2009. Developmental shift in the discrimination of vowel contrasts in bilingual infants: is the distributional account all there is to it? *Dev. Sci.* **12:** 874–887.
61. Sundara, M., L. Polka & F. Genesee. 2006. Language-experience facilitates discrimination of/d-/in monolingual and bilingual acquisition of English. *Cognition* **100:** 369–388.
62. Bosch, L. & M. Ramon-Casas. 2011. Variability in vowel production by bilingual speakers: can input properties hinder the early stabilization of contrastive categories? *J. Phonetics* **39:** 514–526.
63. Sebastián-Gallés, N. 2010. Bilingual language acquisition: where does the difference lie? *Hum. Dev.* **53:** 245–255.
64. Sundara, M. & A. Scutellaro. 2010. Rhythmic distance between languages affects the development of speech perception in bilingual infants. *J. Phonetics* **39:** 513.
65. Fennell, C.T., K. Byers-Heinlein & J.F. Werker. 2007. Using speech sounds to guide word learning: the case of bilingual infants. *Child Dev.* **78:** 1510–1525.
66. Mattock, K., L. Polka, S. Rvachew, *et al.* 2010. The first steps in word learning are easier when the shoes fit: comparing monolingual and bilingual infants. *Dev. Sci.* **13:** 229–243.

67. Fennell, C.T. & K. Byers-Heinlein. 2011. Monolingual and bilingual infants' use of atypical phonetic information in word learning. In *36th Annual Boston University Conference on Language Development*. Boston.
68. Näätänen, R., P. Paavilainen, T. Rinne, *et al.* 2007. The mismatch negativity (MMN) in basic research of central auditory processing: a review. *Clin. Neurophysiol.* **118:** 2544–2590.
69. Shafer, V.L., Y.H. Yu & H. Datta. 2011. The development of English vowel perception in monolingual and bilingual infants: neurophysiological correlates. *J. Phonetics* **39:** 2770.
70. Garcia-Sierra, A., M. Rivera-Gaxiola, C.R. Percaccio, *et al.* 2011. Bilingual language learning: an ERP study relating early brain responses to speech, language input, and later word production. *J. Phonetics* **39:** 546–557.
71. Petitto, L.A., M.S. Berens, I. Kovelman, *et al.* The "Perceptual Wedge" hypothesis as the basis for bilingual babies' phonetic processing advantage: new insights from fNIRS brain imaging. *Brain Lang.* In press.
72. Conboy, B.T. & D.L. Mills. 2006. Two languages, one developing brain: event-related potentials to words in bilingual toddlers. *Dev. Sci.* **9:** F1–12.
73. Marchman, V.A., A. Fernald & N. Hurtado. 2010. How vocabulary size in two languages relates to efficiency in spoken word recognition by young Spanish–English bilinguals. *J. Child Lang.* **37:** 817–840.
74. Ramon-Casas, M., D. Swingley, N. Sebastián-Gallés, *et al.* 2009. Vowel categorization during word recognition in bilingual toddlers. *Cognit. Psychol.* **59:** 96–121.
75. Waxman, S.R. & D.G. Hall. 1993. The development of a linkage between count nouns and object categories: evidence from fifteen-to twenty-one-month-old infants. *Child Dev.* **64:** 1224–1241.
76. Markman, E.M. & G.F. Wachtel. 1988. Children's use of mutual exclusivity to constrain the meanings of words. *Cognit. Psychol.* **20:** 121–157.
77. Halberda, J. 2003. The development of a word-learning strategy. *Cognition* **87:** B23–B34.
78. Markman, E.M. 1992. Constraints on word learning: speculations about their nature, origins, and domain specificity. *Minn. Sym. Child Psych.* **25:** 59–101.
79. Au, T.K., & M. Glusman. 1990. The principle of mutual exclusivity in word learning: to honor or not to honor? *Child Dev.* **61:** 1474–1490.
80. Frank, I., & D. Poulin-Dubois. 2002. Young monolingual and bilingual children's responses to violation of the mutual exclusivity principle. *Int. J. Biling.* **6:** 125–146.
81. Byers-Heinlein, K. & J.F. Werker. 2009. Monolingual, bilingual, trilingual: infants' language experience influences the development of a word-learning heuristic. *Dev. Sci.* **12:** 815–823.
82. Houston-Price, C., Z. Caloghiris & E. Raviglione. 2010. Language experience shapes the development of the mutual exclusivity bias. *Infancy* **15:** 125–150.
83. Emmorey, K., H.B. Borinstein, R. Thompson, *et al.* 2008. Bimodal bilingualism. *Lang. Cognition* **11:** 43–61.
84. Petitto, L.A., M. Katerelos, B.G. Levy, *et al.* 2001. Bilingual signed and spoken language acquisition from birth: implications for the mechanisms underlying early bilingual language acquisition. *J. Child Lang.* **28:** 453–496.
85. Pyers, J.E. & K. Emmorey. 2008. The face of bimodal bilingualism. *Psychol. Sci.* **19:** 531–535.
86. Emmorey, K., G. Luk, J.E. Pyers, *et al.* 2008. The source of enhanced cognitive control in bilinguals. *Psychol. Sci.* **19:** 1201.
87. Bosch, L. & N. Sebastián-Gallés. 1997. Native-language recognition abilities in 4-month-old infants from monolingual and bilingual environments. *Cognition* **65:** 33–69.
88. Byers-Heinlein, K. 2011. Parental language mixing: its measurement and the relation of mixed input to young bilingual children's vocabulary size. *Biling. Lang. Cogn.* In press.
89. Weisleder, A. & A. Fernald. 2011. Richer language experience leads to faster understanding: language input and processing efficiency in diverse groups of low-SES children. In *36th Annual Boston University Conference on Language Development*. Boston.
90. De Houwer, A., M.H. Bornstein & S. De Coster. 2006. Early understanding of two words for the same thing: a CDI study of lexical comprehension in infant bilinguals. *Int. J. Biling.* **10:** 331–347.
91. Pearson, B.Z., S.C. Fernández & D.K. Oller. 1993. Lexical development in bilingual infants and toddlers: comparison to monolingual norms. *Lang. Learn.* **43:** 93–120.
92. Genesee, F., E. Nicoladis & J. Paradis. 1995. Language differentiation in early bilingual development. *J. Child Lang.* **22:** 611–631.
93. Comeau, L., F. Genesee & M. Mendelson. 2007. Bilingual children's repairs of breakdowns in communication. *J. Child Lang.* **34:** 159–174.
94. Gervain J. & J.F. Werker. Unpublished results.

Ann. N.Y. Acad. Sci. ISSN 0077-8923

ANNALS OF THE NEW YORK ACADEMY OF SCIENCES
Issue: *The Year in Cognitive Neuroscience*

# Understanding disgust

Hanah A. Chapman and Adam K. Anderson
Department of Psychology, University of Toronto, Toronto, ON, Canada

Address for correspondence: Hanah Chapman, Division of Humanities and Social Sciences, California Institute of Technology, 1200 E. California Blvd. MC-228-77, Pasadena, CA 91125. hchapman@caltech.edu

**Disgust is characterized by a remarkably diverse set of stimulus triggers, ranging from extremely concrete (bad tastes and disease vectors) to extremely abstract (moral transgressions and those who commit them). This diversity may reflect an expansion of the role of disgust over evolutionary time, from an origin in defending the body against toxicity and disease, through defense against other threats to biological fitness (e.g., incest), to involvement in the selection of suitable interaction partners, by motivating the rejection of individuals who violate social and moral norms. The anterior insula, and to a lesser extent the basal ganglia, are implicated in toxicity- and disease-related forms of disgust, although we argue that insular activation is not exclusive to disgust. It remains unclear whether moral disgust is associated with insular activity. Disgust offers cognitive neuroscientists a unique opportunity to study how an evolutionarily ancient response rooted in the chemical senses has expanded into a uniquely human social cognitive domain; many interesting research avenues remain to be explored.**

**Keywords:** disgust; distaste; morality; emotion; facial expression; interoception

## Introduction

From unsavory foods to squalid restrooms, mutilations to moral depravity, the stimuli that evoke disgust are perhaps the most diverse of any human emotion. Nonetheless, these kinds of objects and events seem to trigger a common experience of revulsion and offence in human beings the world over.[1–3] Disgust has been recognized as a basic and universal human emotion at least since the time of Darwin,[4] but with the exception of pioneering work by Paul Rozin and his colleagues, disgust was largely ignored by the affective revolution that swept through psychology beginning in the 1980s. This trend of neglect has reversed in recent years, however, with an explosion of research on all aspects of disgust. Here we provide a review of recent and classic work on disgust, with an emphasis on what is known about its neural basis. We begin by describing a key evolutionary theory of disgust that makes sense of the heterogeneous assortment of disgusting stimuli, and that provides the conceptual foundation for much of modern disgust research.[1,5]

## Disgust: origins and expansion or descent with modification

Disgust is perhaps best understood by analogy to a phylogenetic or evolutionary tree, with more specialized forms branching off from a root that is the "common ancestor" for all of the varied forms of disgust.[1] The ancestral process for disgust is thought to be distaste, a form of motivated food rejection triggered by the ingestion of unpleasant-tasting substances, prototypically those that are bitter.[1] The behavioral tendency of distaste is oral rejection, that is, spitting out the unpleasant substance. Distaste responses can be seen in adult humans[6] and neonates only a few hours old,[7,8] as well as nonhuman animals, including rats,[9] apes, and monkeys.[10] Because many toxins are bitter,[11] distaste has a clear and concrete adaptive function in motivating the avoidance of poisonous foods. Consistent with this basic adaptive role, the ability to detect and reject bitter substances seems to be very ancient: even sea anemones, which first evolved nearly 500 million years ago, will eject bitter foods from their gastrovascular cavity.[11] Interestingly, the relationship between bitterness and toxicity may

doi: 10.1111/j.1749-6632.2011.06369.x

 © 2012 New York Academy of Sciences.

represent a form of coevolved signaling between predators and prey.[11] Biological toxins likely evoked to defend prey species (including plants) against predation (including grazing). The bitter taste associated with toxins serves to warn to potential predators about the cost of consuming a particular prey species. Both sides benefit from this communication: the predator avoids ingesting more than a mouthful of the toxic prey, and the prey avoids extensive injury.

Although the terms *distaste* and *disgust* are often used interchangeably, the two systems are not identical.[1] In particular, although distaste is focused on the avoidance of toxins,[11] most forms of disgust serve to defend the organism against parasitic disease, including infection by microorganisms such as bacteria and viruses.[12,13] The problem of detecting and avoiding disease is rather more difficult than the problem of detecting and avoiding toxins, because it is in the parasite's interest to infect the host without being detected. Accordingly, parasites usually do not signal their presence in the same way that prey species may signal their toxicity. Instead, an organism that wishes to avoid infection must recognize and avoid stimuli that are reliably—but incidentally—associated with contamination by parasites.[14,15] Relevant cues may include certain types of odors (e.g., the smell of decay), as well as tactile and visual cues (e.g., slime, mold, worms, body products such as feces, certain insects, and sick conspecifics).[12] In species with the cognitive capacity to do so, it is also helpful to avoid objects that may have contacted a primary disease vector—that is, to avoid items that may be contaminated.[16] For example, the clothing or bedding of a person who has a skin infection may be almost as infectious as the diseased person themself.

The psychological features of disgust are consistent with a role in avoiding disease. Compared to distaste, disgust is less reliant on the sense of taste to diagnose potential threats[1] because many modalities can provide information about disease cues. Put more concretely, you do not need to eat a cockroach to be disgusted by it. Disgust also differs from distaste in that disgusting substances are much more contaminating than distasteful substances.[5] Although you may eat around a bitter and disliked vegetable on your plate, you are unlikely to do so if someone spits in your dinner. The property of contamination complements disgust's role in defending against infectious disease: microorganisms in particular can spread invisibly and easily from one substance to another, and a single organism can multiply exponentially to become a serious threat to health.[15] By contrast, toxins tend to be inert and may be harmless if sufficiently diluted.

In spite of the differences between distaste and disgust, the most basic forms of disgust share distaste's behavioral tendency of oral rejection.[1] Accordingly, disgust is thought to have originated from distaste; in other words, disgust is thought to be a branch off the distaste root. Consistent with this descent, the original forms of disgust—often referred to as "core" disgust—are believed to focus on defending against infection via the oral route.[1] From here, other branches developed as the function of disgust expanded to include defense against other types of threats.[1] Still closely tied to disease-avoidance, but less to oral incorporation, is disgust elicited by contact with unfamiliar, unhygienic, or diseased conspecifics, known as interpersonal disgust.[1] Somewhat less directly associated with disease, but still clearly related to biological fitness, are the various forms of sexual disgust.[1] These kinds of disgust may motivate the avoidance of sexual contact with partners who are undesirable from an evolutionary perspective, such as relatives, the very old or very young, and members of the wrong species or the wrong sex.[17] Finally, violations of the normal outer envelope of the body, such as injuries and blood, can also trigger disgust.[1] A number of diseases can spread through contact with blood, so this type of disgust could also serve a disease-avoidance function.[3] Disgust triggered by blood and injuries is sometimes grouped together with sexual disgust to form the category of "animal reminder" disgust, on the logic that these types of stimuli are disturbing because they remind us that humans are mortal animals.[1] Table 1 provides a summary of the stimulus triggers and hypothesized functions for distaste and different types of disgust.

Taken together, we refer to disgust elicited by the rather concrete assortment of stimuli just described as *physical disgust*. We use this umbrella term primarily for convenience, not because of any strong conviction that the various physical disgusts form a homogeneous group. In fact, few studies have directly compared different kinds of physical disgust in an effort to characterize differences and similarities between them.

**Table 1.** Varieties of distaste and disgust

| Type | Example stimulus triggers | Hypothesized function |
|---|---|---|
| Distaste | Unpleasant tastes, especially bitter | Avoid toxins |
| Physical disgust | | |
| Core disgust | Feces, vomit, rats, maggots, spoiled food | Avoid infection via oral route |
| Blood–injury | Injuries, blood, bodily deformities | Avoid infection |
| Interpersonal | Contact with diseased or unfamiliar individuals | Avoid infection |
| Sexual | Sexual contact with very old or very young, wrong sex, or wrong species | Avoid compromising reproductive fitness |
| Moral disgust | Violation of social and moral norms | Avoid unsuitable interaction partners |

Accordingly, it is not clear whether we should expect substantial neural differences between disgust associated with different types of physical stimuli. An exception is research comparing core disgust to disgust triggered by blood and injuries (BI disgust). Core and BI disgust are associated with different psychophysiological correlates: core disgust is related to nausea and changes in the normal rhythm of stomach contractions,[18,19] whereas BI disgust is related to light-headedness and fainting, associated with changes in the cardiovascular system.[18,20] Core and BI disgust are also associated with different clinical phenomena: core disgust with OCD symptoms,[21] and BI disgust with blood–injection–injury phobia.[22,23]

Although distaste can evidently be seen in many nonhuman animals, it is less clear whether physical disgust exists beyond our own species.[5] On the one hand, many nonhuman animals do not seem to show the same aversion to disease vectors (e.g., feces) that humans do.[5] On the other hand, many species *do* show clear evidence of a variety of disease-avoidance behaviors.[24] It is not clear whether such behavior is accompanied by a subjective experience of disgust that is similar to what humans feel; nonetheless, some accompanying motivational state seems likely, and this may represent disgust in nonhuman animals.

Some of the confusion over disgust in nonhuman animals may be related to the different levels of disease risk faced by different species. In particular, some species may be more vulnerable to disease than others, and hence may have a greater need for disgust. For example, humans are omnivores: this lack of dietary specialization may lead to frequent exposure to dangerous foodstuffs.[5] By contrast, species with a highly specialized diet may simply never come into contact with foods that may carry disease. At the other end of the spectrum, species adapted for scavenging may have developed specialized mechanisms to deal with the challenges of eating rotten food. Humans' highly social lifestyle may also have encouraged the evolution of a sensitive disgust system: one of the costs of social life is increased risk of exposure to disease from conspecifics.[5,25] Animal species that are asocial, or that live in relatively isolated groups, may not need mechanisms to mitigate this risk.

There is one form of disgust that does seem unique to humans, namely disgust triggered by the violation of social norms and moral values.[1] For example, people who steal, lie, cheat, and harm others are all referred to as "disgusting."[3,26,27] Disgust's leap from the physical world of disease avoidance to the much more abstract sociomoral domain is quite striking, and may represent an example of exaptation,[1] an evolutionary process whereby a pre-existing system assumes a new functional role.[28–30] In the case of moral disgust, the new functional role may be motivating the avoidance of individuals who violate social norms, who accordingly may not be good partners for interaction.[1] It remains unclear whether there are particular types of moral transgressions that are most strongly tied to disgust.[31,32]

Although anger may seem like a more natural response to norm violations than disgust, it is worth considering that anger is an approach-related, strongly activating emotion.[33] Hence, it may represent a rather costly response to moral transgressions. By contrast, the withdrawal and avoidance motivation associated with disgust may offer a lower-cost strategy.[32] Indeed, recent modeling work suggests that noncooperation is often a more efficient response to norm violation than is costly punishment.[34] That said, moral disgust is not without controversy:[35] some have argued that moral disgust may just be anger in disguise,[36] or that moral disgust may be limited to transgressions that remind us of physical disgust stimuli (e.g., gory murders).[13,37] We discuss the debate surrounding moral disgust in more detail below.

To summarize, disgust is believed to have expanded from an origin in distaste and the avoidance of toxins, through to avoidance of disease and other threats to biological fitness, and finally into the social and moral domain (Table 1).[1,5] The broad scope of disgust provides fertile ground for a number of cognitive neuroscience research questions, and in turn, cognitive neuroscience provides new avenues to test this theory of the evolution of disgust. A key question is whether the different forms of disgust are related to one another at the neural level, and on the flip side, how their differing functional roles are instantiated. Cognitive neuroscience research on disgust first began more than a decade ago, with work examining the neural basis of perceiving disgusted facial expressions.[38–40] More recent work continues to be heavily influenced by these early findings, so we will start our review of the cognitive neuroscience of disgust by considering the neural correlates of perceiving disgust expressions before going on to discuss distaste, physical disgust, and moral disgust. As we will see, the insula has been strongly implicated in perceiving as well as experiencing many forms of disgust. We therefore begin by providing a short overview of what is known about the insula.

## Insular cortex: anatomy and function

Concealed beneath the overlying frontal, parietal, and temporal opercula, the human insula consists of five to seven gyri with substantial morphological variation between individuals (Fig. 1).[41,42] It is interconnected with a number of cortical regions, including anterior cingulate, frontal pole, and dorsolateral prefrontal cortex as well as primary and secondary somatosensory cortex and auditory cortex. The insula is also heavily connected with the amygdala and dorsal thalamus.[43–45]

Architectonic studies have revealed three major divisions of insula in most mammals: an anterior agranular section, a middle dysgranular section, and a posterior granular section.[46] In humans and great apes, there is an additional sector in the most anterior and ventral portion of the insula, at its junction with the orbitofrontal cortex.[47] This area, known as frontoinsular cortex, contains the distinctive von Economo neurons (VENs),[48] large bipolar neurons that are especially prominent in humans and great apes but not other primates.[47] VENs are selectively compromised in early-stage frontotemporal dementia, in which empathy, social awareness, and self-control deteriorate.[49]

The insula has been known to play a role in viscerosensation, olfaction, and gustation since Penfield's experiments.[50] Modern research has confirmed the important interoceptive function of the insula, and also suggests a posterior-to-mid-to-anterior functional gradient, from primary interoceptive representations in posterior insula, to a middle integration zone, and finally to an anterior region involved in high-level integration of homeostatic information with other cognitive and affective processes.[51] Indeed, the connectivity of the insula places it in an ideal position to integrate homeostatic information with information about the physical and social external environment. The role of the insula also seems to extend well beyond interoception, to include a wide variety of cognitive, affective, and social processes.[51] The sheer breadth of processes associated with insular activation, as well as the effects of VEN degeneration in frontotemporal dementia, have led to the suggestion that the insula, especially its most anterior sections, may underlie the human sense of self-awareness or consciousness.[51]

## Perceiving disgusted facial expressions

Having completed this brief review of the insula, we now return to the cognitive neuroscience of disgust, beginning with what is known about perceiving disgust in others. Darwin was perhaps the first to recognize and describe a distinct facial expression associated with disgust.[4] The canonical disgust expression centers around movements of the mouth and nose, including raising of the upper lip and wrinkling of the nose.[4,52,53] Gape-like opening of the mouth may also be present, as well as lowering of the brows.[1] In addition to serving as a social signal, the disgust expression may provide egocentric benefits to the sender. In particular, some of the facial movements of disgust may serve to defend

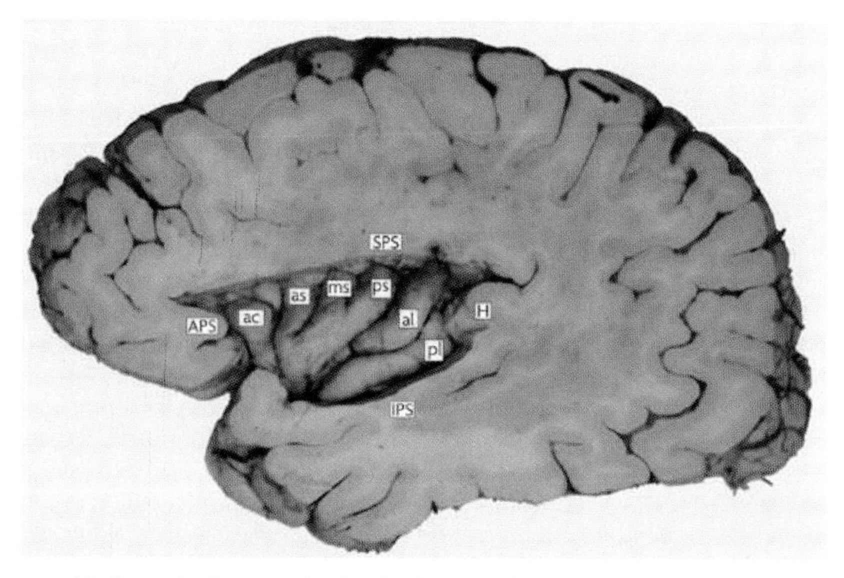


**Figure 1.** Anatomy of the human insula. as, anterior short insular gyrus; al, anterior long insular gyrus; ac, accessory gyrus; APS, anterior peri-insular sulcus; H, Heschl's gyrus; IPS, inferior peri-insular sulcus; ms, middle short insular gyrus; ps, posterior short insular gyrus; pl, posterior long insular gyrus; SPS, superior peri-insular sulcus. Photograph is courtesy of Profs. A.D. Craig and Thomas Naidich;[51] reproduced with permission from Nature Publishing Group.

the vulnerable mucous membranes of the eyes and nose from contact with contaminants.[1,54] Wrinkling the nose and raising the upper lip have the effect of decreasing the volume of the nasal cavities and reducing the amount of air that is inhaled through the nose.[54] Similarly, lowering the brows decreases the exposed surface of the eyes.[54] By contrast, the mouth opening that is sometimes seen in disgust may ready the individual to spit or vomit out any already-ingested food,[1] and increased salivation may help to flush contaminated material from the mouth.[1,55]

The canonical disgust expression is recognized cross-culturally, leading to the suggestion that disgust is one of the basic and universal human emotions.[56,57] Moreover, continuity of the upper lip raise across distaste, physical disgust, and moral disgust provides some of the only empirical evidence for the evolutionary expansion of disgust described above.[58] That said, there is debate as to how similar the expressions associated with distaste and the different forms of disgust really are.[59,60] One study has suggested that mouth gaping may be most strongly associated with distaste, whereas the upper lip raise may be more closely tied to forms of disgust that are removed from oral rejection (e.g., moral disgust).[59] However, this research examined recognition of posed facial expressions rather than measuring spontaneous expression production, and the validity of this approach for examining subtle differences in expression configuration is not clear. Moreover, only the canonical disgust expression, including the upper lip raise and/or nose wrinkle, is associated with activation of the anterior insula, which we will shortly see has been tied to many forms of disgust.[60]

In spite of this debate, the majority of neuroimaging research on disgust expression perception has not been concerned with potential differences among different types of disgust expressions. Rather, the most common question has been whether there are distinct neural substrates for perceiving different emotional expressions (e.g., disgust, fear, anger, etc.). The initial work on this question generated evidence that viewing the canonical disgust expression is associated with activation of the anterior insula, relative to viewing neutral facial expressions.[38–40,61] Although insular activation was also observed for fearful expressions, the insula responded most strongly to disgust when the two were directly compared (but see also Ref. 62). Consistent with these neuroimaging findings, one intracranial recording study in human epilepsy patients implanted with insular depth electrodes found a number of contacts that responded more strongly to facial expressions of disgust than to other emotions.[63]

A recent meta-analysis of 105 fMRI studies of facial expression perception confirms the association between disgust expression perception and activation of the anterior insula.[64] Angry expressions were also found to result in insular activation, but a direct comparison of the activation likelihood estimates for disgust and anger revealed a greater likelihood of activation for disgust.[64] There is some evidence that the anterior insula may provide information about disgust to other brain structures: under conditions of divided attention to disgust expressions, there is a reciprocal relationship between insular activity (reduced with divided attention) and amygdala activity (increased with divided attention).[61] Because the amygdala responds only to fearful expressions under full attention, information from insula may influence the breadth/narrowness of amygdala response tuning.[61]

An interesting extension of expression perception research is work that has examined whether the same brain regions are activated both when an individual perceives a disgusted facial expression and when they personally experience disgust. One study reported activation of the same regions of anterior insula both when subjects viewed videos of actors expressing disgust after smelling an unpleasant odor and when subjects personally experienced unpleasant odors delivered via an olfactometer.[65] Similarly, overlapping regions of anterior insula were activated when subjects viewed videos of actors tasting unpleasant liquids, when they personally tasted unpleasant liquids, and when they imagined physically disgusting events.[66] The results of these studies are important in the wider field of affective psychology and neuroscience, because they provide support for embodied or simulationist theories of facial expression perception. These theories argue that we understand the facial expressions, and indeed the emotions of others, by activating a similar emotion in ourselves.[67,68] Evidence for neural overlap between the perception of disgust expressions and the experience of disgust thus provides an important contribution to our understanding of emotion, empathy, and social communication.

To our knowledge, there have been no human neuroimaging studies of disgust expression production (as opposed to perception). However, a recent intracranial stimulation study in monkeys found that stimulation of ventral anterior insula resulted in facial grimacing, including curling of the upper lip and wrinkling of the nose, that resembled spontaneous responses to unpleasant stimuli.[69] These results suggest that anterior insula may participate in the production as well as the perception of disgust expressions. That said, a study of human patients implanted with insular depth electrodes did not report production of disgust expressions in response to stimulation of the ventral anterior insula, although two patients did report unpleasant sensations in the mouth and throat.[63]

Although the evidence just reviewed emphasizes the importance of the insula in perceiving disgust facial expressions, the basal ganglia have also been implicated, albeit less consistently. The early fMRI studies of disgust expression perception reported activation of the caudate, putamen, and globus pallidus in response to disgust facial expressions;[38–40,61] however, meta-analysis has not supported these findings.[64] There have been several reports of a specific impairment in recognizing disgust expressions in patients with disorders affecting the basal ganglia, such as Huntington's Disease (HD)[70,71] and Obsessive-Compulsive Disorder (OCD).[72] However, more recent results have been mixed. Some studies have replicated the early findings,[73,74] but others have failed to detect any deficit in disgust expression perception,[75] or have reported more general impairments in perceiving many emotional expressions.[76–78] One case report of a patient with a selective lesion of left basal ganglia and insula described a highly specific impairment in recognizing disgust expressions.[79] However, a comparable patient with a right-hemisphere lesion showed no deficit in disgust expression perception.[80] Thus, the involvement of the basal ganglia in disgust expression perception remains unclear.

In summary, neuroimaging work has strongly implicated the insula, particularly its anterior sectors, in perceiving disgust expressions. A caveat worth noting is that neuroimaging cannot demonstrate whether the insula is *necessary* for perceiving disgust. Given the sparse and mixed lesion evidence available to date, the question of necessity remains an open one.

## Distaste

We now turn from perceiving disgust in others to considering the various forms of subjectively experienced disgust. We begin by examining the neural substrates of distaste, given its apparent status as the precursor of disgust. As mentioned, the distaste response is strongly tied to gustatory sensation. Accordingly, distaste begins in the mouth with stimulation of taste receptor cells. Information from taste receptors is transmitted to the central nervous system via cranial nerves VII, IX, and X.[81] These projections synapse in the nucleus of the solitary tract and then continue to gustatory thalamus.[82] Thalamic efferents project to the anterior insula and overlying operculum,[83,84] whereas a second, less extensive projection terminates in postcentral gyrus.[85] The predominance of the insular projection suggests that the insula likely contains primary gustatory cortex.[86] However, the exact location of human primary gustatory cortex in the insula remains a topic of debate, and there may be multiple gustatory representations in human insula.[86] Some of the uncertainty likely stems from the fact that insular activity is by no means exclusive to taste: rather, the insula responds to a variety of inputs associated with feeding, including oral somatosensation,[87] olfaction,[88] and temperature.[89] Because it is difficult to control for all of these variables in human neuroimaging research, it remains challenging to pinpoint human primary gustatory cortex.[86]

Although the insula is evidently a multimodal region, there is some evidence for functional specialization both in the insula and in other taste-sensitive brain regions. In humans, the dorsal middle insula, as well as the amygdala, seem to respond to taste intensity, irrespective of pleasant or unpleasant valence.[90] By contrast, dorsal anterior insula and some regions of anterior obitofrontal cortex (OFC) have been found to respond preferentially to unpleasant (i.e., distasteful) stimuli, regardless of intensity.[90] In turn, pleasant tastes may be associated with activation of the insula more ventrally and OFC more laterally.[90] An interesting complement to these human neuroimaging findings is an intracranial stimulation study in macaques, which found that stimulation of the ventral anterior insula resulted in spitting out or throwing away preferred foods, as if they had become distasteful.[69] Single-neuron recordings from macaques also provide evidence that distinct cells in the anterior insula respond most strongly to particular pleasant and unpleasant tastes, but do not suggest a topographic organization of these cells (i.e., the location of a taste cell does not predict what taste it responds to).[91–93]

The take-home message here is that although the insula almost certainly has a way of representing distaste, it is clearly not—as a whole—a region that is selective for distasteful stimuli. Rather, the insula responds to both pleasant and unpleasant tastes, as well as other feeding-related stimuli, and likely codes for specific gustatory experiences in a distributed fashion.[91]

## Physical disgust

In contrast to distaste, which is closely tied to gustatory sensation, physical disgust can be evoked by a much wider range of stimuli and via all sensory modalities. Cognitive neuroscientists have studied physical disgust using photographs,[94,95] films,[18,96,97] auditory stimuli,[39] autobiographical recall,[98] script-driven imagery,[66] and written vignettes.[99,100] A wide range of physical disgust stimuli have also been examined, most commonly including body products, such as feces and vomit; spoiled food; insects, such as roaches and worms; blood; injuries; and corpses. Although the stimulus triggers for physical disgust are quite distinct from the simple chemosensory stimuli used to study distaste, physical disgust stimuli are also strongly associated with activity in the anterior insula. For example, viewing disgusting photographs[94,95] and films,[18] imagining disgusting events,[66] and recalling disgusting experiences[98] have all been found to result in anterior insula activation, relative to neutral comparison conditions. Subjective ratings of disgust—but not fear—in response to disgusting photographs are correlated with activation of the anterior insula.[101] Similarly, subjective ratings of disgust in response to disgusting films are correlated with anterior insula activation.[18] Individual differences in the tendency to experience physical disgust (i.e., trait physical disgust) are also associated with the degree of activation of the anterior insula while viewing disgusting photographs.[102,103] That said, individual differences in trait anxiety are also associated with activation of the insula when subjects view fearful photographs.[102]

Although there have been occasional failures to replicate insular activation in response to physical disgust stimuli,[104,105] the association between bilateral insular activation and physical disgust has been confirmed in a recent meta-analysis of 83 neuroimaging studies of emotion, including 29 studies of physical disgust.[106] Smaller clusters of insular activity were also seen for happiness, fear, and sadness;

however, these activations were much smaller than for disgust, and the activation likelihood estimates for disgust were significantly stronger than for happiness, fear, or sadness. Naturally, the insula is not the only brain region associated with disgust. It is not uncommon to see amygdala activity in response to physical disgust, as well as the anterior cingulate and medial fontal gyrus.[106] Activation of these regions may represent nonspecific aspects of disgust experience, such as emotional arousal, withdrawal tendencies, and self-regulation.

The triggers for physical disgust encompass a remarkably diverse set of stimuli, and an obvious question is whether different subtypes of physical disgust stimuli are associated with distinct neural correlates in the insula or in other brain regions. As discussed above, there have been few empirical studies of behavioral and/or physiological differences and similarities between different types of physical disgust; thus, neuroimaging comparisons tend to be exploratory rather than hypothesis driven. An exception is the small handful of studies that have investigated differences between disgust evoked by stimuli such as body products, rotten food, and insects (e.g., core disgust) and disgust evoked by blood and violations of the outer body envelope (e.g., BI disgust). Core and BI disgust are known to be associated with different physiological responses[18,19,20] as well as with different clinical phenomena.[21,22,23] In spite of these differences, core and BI disgust seem to result in partly overlapping activation in the anterior insula, perhaps reflecting a common disgust experience.[18,95,97,107] There are also differences in insular activation between core and BI disgust, with core disgust activating a ventral anterior region more strongly, and BI disgust activating a more dorsal mid-insular area. The characteristic psychophysiological effects of core and BI disgust on the stomach and cardiovascular system, respectively, are also represented differentially in the insula.[18] Differences between core and BI disgust have also been reported in other brain regions, although the exact areas have been somewhat inconsistent across studies, perhaps due to methodological differences.[18,95,97,107]

Less is known about the neural correlates of sexual disgust as compared to other forms of physical disgust. One study has contrasted the neural responses to written scenarios describing incest and scenarios describing core disgust stimuli.[100] The authors found stronger activation of the anterior insula in response to the incest relative to the core disgust stimuli.[100] However, this study is somewhat unusual in that it did not find significant insular activation in response to the core disgust stimuli compared to neutral, perhaps because written stimuli were used, as compared to the films or photographs used in most neuroimaging studies of physical disgust. The only other study to investigate sexual disgust did use photographs as stimuli, but found no insular activation to either sexual disgust photographs or core disgust photographs, compared to neutral photographs.[108] This latter study comes from a group that has performed a number of neuroimaging studies of physical disgust, sometimes finding insular activity[94,101,109] and other times not.[104,107,108] Accordingly, it is somewhat difficult to interpret their null results on insular activation in response to sexual disgust stimuli; indeed, it is difficult to interpret null results in any neuroimaging study. The relationship between sexual disgust and other forms of physical disgust thus remains an open avenue for further exploration. It seems likely to be an interesting one, given recent findings that tie disapproval of homosexuality to trait physical disgust and political conservatism.[110,111]

A final question is whether physical disgust is indeed related to its supposed precursor, distaste, at the neural level. To our knowledge, only one study has directly compared the neural correlates of tasting unpleasant liquids and experiencing physical disgust.[66] To induce distaste, subjects drank bitter quinine solutions, whereas to induce physical disgust, subjects imagined disgusting incidents (e.g., ingesting vomit). A common region of anterior insula/frontal operculum was active in both conditions, suggesting that physical disgust and distaste are indeed supported by at least partially overlapping neural substrates.

## Is insular activation specific to disgust?

The results described so far suggest an important role for the insula, especially its anterior regions, in perceiving disgust expressions as well as experiencing physical disgust and distaste. Moreover, there is some evidence that strong activation of the insula may be particularly characteristic of disgust: the two recent meta-analyses described above, one concentrating on facial expressions[64] and the other including a wider range of emotional stimuli,[106] both showed stronger activation of the insula for disgust as compared to other emotional conditions. In spite of these findings, however, it seems unlikely that there is a unique, one-to-one mapping between activation of the anterior insula and disgust. Some of the results described above begin to hint at this: for example, the insula responds to a wide range of feeding-related stimuli beyond distaste, including oral somatosensation,[91] olfaction,[88] and ingestive motor activity.[86,91] Beyond this, the insula, including its anterior regions, responds to a variety of interoceptive stimuli including heartbeat,[112] stomach and bladder distention,[113,114] sexual arousal,[115] and itch,[116] among many others.[51] Lest researchers of a more cognitive bent feel left out, the insula is also activated by a variety of more traditionally cognitive phenomena, including goal-directed attention,[117] cognitive control and performance monitoring,[118] risk and uncertainty,[119,120] and perceptual decision making.[51,121]

Findings such as these lead us to believe that although the insula may be very important in disgust and distaste, it is not specific to these states. This position may initially seem difficult to reconcile with the meta-analytic results pointing to at least some degree of insular selectivity for disgust.[64,106] We suggest two possible explanations for the discrepancy. One possibility is that among the emotions, disgust may be particularly strongly associated with visceral changes, consistent with its apparent origins in defending against ingestion of toxic or contaminated foods. Indeed, taste itself may represent an interoceptive rather than an exteroceptive sense.[122] Given that the insula seems to play a key role in interoception,[123] heightened insular activity in response to disgust may simply reflect disgust's strong visceral component.

A second, related, but perhaps more mundane possibility is that disgust *per se* may not have a stronger visceral component than other emotions. Rather, the stimuli that are used to evoke disgust in the laboratory may simply be more effective at causing visceral changes than the stimuli that are used to evoke other emotions. On this view, more effective stimulus triggers for other emotions could also result in insular activity. This latter possibility could be tested by using more compelling comparison stimuli as a contrast to disgust.

We wish to note that our arguments against insular specificity to disgust are directed against a more traditional, "locationist" notion of specificity. In other words, what we disagree with is the idea that the insula as a whole, or even a specific region of it, is uniquely activated by disgust. We do believe that the insula must have a way of representing disgust and distaste as states that are distinct from other emotional and motivational experiences. However, this representation is likely a distributed one within the insula and may well extend beyond the insula. We also do not mean to suggest that past research linking the insula more broadly to disgust is uninformative; rather, it serves to highlight the insula as a region to focus on in more targeted analyses.

## Sociomoral disgust

Most researchers will probably find the received view of how physical disgust evolved to be quite plausible,[1] and many may also accept an important—although perhaps not unique—role for the insula in distaste and physical disgust. Understandably, however, some may be more skeptical of sociomoral disgust. Is sociomoral "disgust" really related to more basic forms of disgust, or is it just a compelling metaphor used to condemn antisocial behavior?[13,124] Even if one can accept that sociomoral disgust is derived from physical disgust, maybe it is only triggered by transgressions that contain reminders of physical disgust, such as bloody murders and depraved sexual crimes?[37]

These questions remain unresolved in the behavioral literature. On the one hand, there is evidence that sociomoral disgust is indeed a genuine form of disgust, and that it is not limited to transgressions that involve physical disgust stimuli. For example, both adults and children call moral transgressions disgusting,[26,32,125] and match them to disgusted facial expressions,[36,58,125] even when the transgressions do not reference physical disgust. Moral transgressions result in raising of the upper lip,[58,126] a characteristic element of the disgust expression, as do physical disgust and distaste.[58] The upper lip raise in response to transgressions is correlated with self-reported disgust, but not anger or contempt.[58] Individuals who are higher in trait physical disgust make more severe judgments about moral transgressions than do their low-physical-disgust counterparts,[127] an effect that is not accounted for by more general differences in trait negative affect. Finally, experimentally induced distaste[128] and physical disgust[129,130] cause changes in moral judgments about issues that do not concern physical disgust.

On the other hand, concerns have been raised about some of these findings,[35,131] and conflicting evidence also exists. For example, moral disgust seems to be relatively insensitive to the intent of a perpetrator and the degree of harm resulting from a transgression,[37,132] factors that are generally considered to be important in moral reasoning. Experimentally induced disgust and individual differences in trait disgust may also have a stronger influence on moral judgments about transgressions that contain reminders of physical disgust stimuli, such as sexually promiscuous behavior, relative to transgressions that contain no such reminders.[133] Measurement issues could explain some of the divergent results: self-reports of moral disgust seem to be quite sensitive to how questionnaire measures are phrased.[32,134] Similarly, different studies use quite dramatically different "moral" stimuli, which could have distinct relationships to moral disgust.

Moral disgust thus remains a contentious topic. Neuroimaging could potentially inform the debate by revealing whether moral and physical disgust have a common neural substrate. However, neuroimaging studies of moral disgust present their own challenges. For example, imagine a case where anterior insula activity was detected when subjects made judgments about moral transgressions (for example, Ref. 135). Could this represent sociomoral disgust, and provide evidence that moral disgust is indeed related to physical disgust? Given that the insula is not a unique substrate for disgust,[51] this is a rather questionable reverse inference.[136]

The inference would be somewhat stronger if a direct comparison between sociomoral and physical disgust in the same subjects revealed common insular activity. A handful of studies have now used this kind of design. However, two of these are somewhat difficult to interpret, as they did not find insular activation in response to the physical disgust conditions, perhaps because written rather than visual stimuli were used.[99,100] The only other study did find insular activation in response to a sexual disgust condition (even though written vignettes were used).[137] However, this study did not find insular activity in response to purely moral stimuli, such as descriptions of harm and dishonesty, relative to neutral stimuli. Once again though, the null result for sociomoral disgust is difficult to interpret. The disgusting sexual scenarios were rated as more emotionally arousing than the sociomoral stimuli;[137] it is possible that more arousing sociomoral transgressions could have resulted in activation of the anterior insula. Indeed, very difficult and emotional moral dilemmas, such as whether to kill a crying baby to save an entire village from enemy soldiers, have been found to cause activation of the anterior insula.[135] Similarly, actions that are judged to be morally wrong, as well as controversial moral transgressions, result in increased insular activation relative to actions that are judged to be not wrong and noncontroversial transgressions.[138]

These findings suggest that future work could potentially reveal overlapping activation of the anterior insula for both moral transgressions and physical disgust. However, any such overlap could still represent more generic similarities between moral cognition and physical disgust, such as emotional arousal or uncertainty, rather than a shared disgust experience *per se*. Accordingly, it is likely that neuroimaging studies will provide converging, rather than conclusive, evidence in the debate over moral disgust. It will also be very important for future studies in this area to use moral and physical disgust stimuli that are carefully controlled for spurious differences on dimensions, such as emotional arousal, so as not to confound comparisons of interest.

If moral disgust does indeed prove to be related to physical disgust, then neural differences between moral and physical disgust may also be very interesting. For example, moral disgust is triggered by much more abstract and social stimuli than most forms of physical disgust. Accordingly, moral disgust may involve brain regions that play a role in social cognitive processing, such as dorsal medial prefrontal cortex[139] and temporoparietal junction,[140,141] as well as regions involved in more basic forms of disgust such as insula and perhaps basal ganglia.

## Outstanding questions and future directions

Behavioral and cognitive neuroscience studies have considerably increased our understanding of disgust—in all its forms—in recent years. In particular, there is now evidence that distaste and various forms of physical disgust are indeed related to one another at the neural as well as the behavioral levels, as proposed by Paul Rozin and his colleagues many years ago.[1,5] Nonetheless, relative to other emotions such as fear and sadness, the study of disgust remains in its infancy. One area in particular need of further study is the degree of similarity between different forms of disgust. For example, an influential typology of disgust groups BI disgust and sexual disgust together to form the category of "animal reminder" disgust, which is believed to be rooted in anxiety over human mortality.[1] Neuroimaging studies could inform our understanding of the structure of disgust by revealing commonalities and differences in the neural processing of various disgust stimuli. These kinds of studies may be especially valuable for moral disgust, whose relationship to physical disgust remains a topic of intense debate.[35,36,58]

Neuroimaging comparisons of different types of disgust would likely benefit from the application of more advanced techniques, such as connectivity analyses and multivoxel pattern analysis, which afford analysis of the distributed representations supported by the anterior insula. Beyond functional neuroimaging, electroencephalography (EEG) may also prove a useful tool, especially for investigating sociomoral disgust. In particular, although anger is closely associated with approach motivation and increased EEG alpha power in the left relative to right hemispheres,[33] disgust is believed to be associated with withdrawal motivation[1] and perhaps increased alpha power in the right hemisphere.[142] This raises the possibility that sociomoral disgust may also be associated with increased right hemisphere alpha power and with a stronger tendency to withdraw from rather than approach transgressors.[32]

Another potential research direction is to examine the neural substrates underlying the perception that an object is contaminated or contaminating, a property that is unique to disgusting stimuli.[1,5] To our knowledge, no neuroimaging study has yet investigated contamination. Improved understanding of contamination could have important clinical implications, because excessive contamination concerns feature prominently in some forms of obsessive-compulsive disorder.[143] Speculatively, contamination may involve a memory component, because history of contact is the defining feature of contamination. In this way, the study of contamination could form a bridge between disgust research and the wider field of memory research. Similarly, there is a small but growing literature on the cognitive effects of disgust (for example, Refs. 144–146), but no studies have yet examined the influence of disgust on such processes as attention and memory at the neural level.

A final, and rather different direction is to examine the potential relationship between distaste, disgust, and pain. For example, concentrated bitter solutions can produce a painful burning sensation and result in activation of the trigeminal nerve, which carries nociceptive information to the central nervous system.[147,148] In our own research, we have found that the facial expression produced in response to distaste closely resembles the facial response to pain.[149] Moreover, BI disgust is, by definition, closely tied to the perception of physical injury. How are pain and disgust related at the neural level? Is blood-injury disgust more closely related to pain than core disgust? How does pain influence disgust, and vice versa? Finally, how does disgust toward another's injuries influence empathetic responses to their suffering? These questions remain to be examined.

## Conclusion

Scientific knowledge of disgust has expanded far beyond what Darwin might have foreseen when he provided the first empirical description of this emotion more than one hundred years ago. Recent research has provided important support for classic theoretical work, especially for the idea that specialized forms of disgust—including core, BI, and perhaps also sociomoral disgust—are related to one another and descended from distaste, an ancient motivational response rooted in the chemical senses.[1] Exciting new directions have also emerged, including functional neuroimaging work that has increased not only our understanding of disgust, but also of empathy, social cognition, and emotion, more broadly. We are confident that many more promising research directions exist, to be suggested by future findings and creative research yet to be done.

## Conflicts of interest

The authors declare no conflicts of interest.

## References

1. Rozin, P., J. Haidt & C. McCauley. 2000. Disgust. In *Handbook of Emotions*. M. Lewis & J. Haviland-Jones, Eds.: 637–653. Guilford Press. New York.
2. Haidt, J., P. Rozin, C. McCauley & S. Imada. 1997. Body, psyche, and culture: the relationship between disgust and morality. *Psychol. Develop. Soc.* **9:** 107–131.

3. Curtis, V. & A. Biran. 2001. Dirt, disgust, and disease. Is hygiene in our genes? *Perspect. Biol. Med.* **44:** 17–31.
4. Darwin, C. 1872/1998. *The Expression of the Emotions in Man and Animals*. Oxford University Press. Oxford.
5. Rozin, P. & A. Fallon. 1987. A perspective on disgust. *Psychol. Rev.* **94:** 23–41.
6. Greimel, E., M. Macht, E. Krumhuber & H. Ellgring. 2006. Facial and affective reactions to tastes and their modulation by sadness and joy. *Physiology and Behavior* **89:** 261–269.
7. Peiper, A. 1963. *Cerebral Function in Infancy and Childhood*. Consultants Bureau. New York.
8. Steiner, J. 1973. The gustofacial response: observation on normal and anencephalic newborn infants. In *Fourth Symposium on Oral Sensation and Perception*. J.F. Bosma, Ed.: 254–278. U.S. Department of Health, Education and Welfare. Bethesda, MD.
9. Grill, H. & R. Norgren. 1978. The taste reactivity test. I. Mimetic responses to gustatory simuli in neurologically normal rats. *Brain Res.* **143:** 263–279.
10. Berridge, K. 2000. Measuring hedonic impact in animals and infants: microstructure of affective taste reactivity patterns. *Neurosci. Biobehav. Rev.* **24:** 173–198.
11. Garcia, J., W.G. Hankins, D.A. Denton & J.P. Coghlan. 1975. The evolution of bitter and the acquisition of toxiphobia. In *Olfaction and Taste*, Vol. V.D.A. Denton & J.P. Coghlan, Eds.: 39–45. Academic Press. New York.
12. Curtis, V., R. Aunger & T. Rabie. 2004. Evidence that disgust evolved to protect from risk of disease. *Proc. R. Soc. B.: Biol. Sci.* **271**(Suppl 4): S131–133.
13. Oaten, M., R. Stevenson & T. Case. 2009. Disgust as a disease-avoidance mechanism. *Psychol. Bull.* **135:** 303–321.
14. Schaller, M. & J.H. Park. 2011. The behavioral immune system (and why it matters). *Curr. Direct. Psychol. Sci.* **20:** 99–103.
15. Pinker, S. 1997. *How the Mind Works*. WW Norton. New York.
16. Rozin, P., L. Millman & C. Nemeroff. 1986. Operation of the laws of sympathetic magic in disgust and other domains. *J. Personal. Soc. Psychol.* **50:** 703–712.
17. Fessler, D.M.T. & C.D. Navarrete. 2003. Domain-specific variation in disgust sensitivity across the menstrual cycle. *Evol. Hum. Behav.* **24:** 406–417.
18. Harrison, N.A., M.A. Gray, P.J. Gianaros & H.D. Critchley. 2010. The embodiment of emotional feelings in the brain. *J. Neurosci.* **30:** 12878.
19. Stern, R.M., M.D. Jokerst, M.E. Levine & K.L. Koch. 2001. The stomach's response to unappetizing food: cephalic-vagal effects on gastric myoelectric activity. *Neurogastroenterol. Motility* **13:** 151–154.
20. Cisler, J., B.O. Olatunji & J. Lohr. 2009. Disgust, fear, and the anxiety disorders: a critical review. *Clin. Psychol. Rev.* **29:** 34–46.
21. Olatunji, B.O., N.L. Williams, D.F. Tolin, *et al.* 2007. The disgust scale: item analysis, factor structure, and suggestions for refinement. *Psychol. Assess.* **19:** 281–297.
22. Olatunji, B.O., K.M. Connolly & B. David. 2008. Behavioral avoidance and self-reported fainting symptoms in blood/injury fearful individuals: an experimental test of disgust domain specificity. *Journal of Anxiety Disorders* **22:** 837–848.
23. Olatunji, B.O., C.N. Sawchuk, P.J. de Jong & J.M. Lohr. 2006. The structural relation between disgust sensitivity and blood-injection-injury fears: a cross-cultural comparison of U.S. and Dutch data. *J. Behav. Ther. Exp. Psychiatr.* **37:** 16–29.
24. Hart, B.L. 1990. Behavioral adaptations to pathogens and parasites: five strategies. *Neurosci. Biobehav. Rev.* **14:** 273–294.
25. Cote, I.M. & R. Poulinb. 1995. Parasitism and group size in social animals: a meta-analysis. *Behav. Ecol.* **6:** 159–165.
26. Simpson, J., S. Carter, S.H. Anthony & P.G. Overton. 2006. Is disgust a homogeneous emotion? *Motivat. Emot.* **30:** 31–41.
27. Tybur, J.M., D. Lieberman & V. Griskevicius. 2009. Microbes, mating, and morality: individual differences in three functional domains of disgust. *J. Personal. Soc. Psychol.* **97:** 103–122.
28. Bock, W. 1959. Preadaptation and multiple evolutionary pathways. *Evolution* **13:** 194–211.
29. Mayr, E. & S. Tax. 1960. The emergence of evolutionary novelties. In *Evolution After Darwin*. Vol. 1: The Evolution of Life. S. Tax, Ed.: 349–380. Univ. Chicago Press. Chicago.
30. Gould, S. & E. Vrba. 1982. Exaptation: a missing term in the science of form. *Paleobiology* **8:** 4–15.
31. Rozin, P., L. Lowery, S. Imada & J. Haidt. 1999. The CAD triad hypothesis: a mapping between three moral emotions (contempt, anger, disgust) and three moral codes (community, autonomy, divinity). *J. Personal. Soc. Psychol.* **76:** 574–586.
32. Hutcherson, C.A. & J.J. Gross. 2011. The moral emotions: a social-functionalist account of anger, disgust, and contempt. *J. Personal. Soc. Psychol.* **100:** 719–737.
33. Carver, C.S. & E. Harmon-Jones. 2009. Anger is an approach-related affect: evidence and implications. *Psychol. Bull.* **135:** 183–204.
34. Ohtsuki, H., Y. Iwasa & M.A. Nowak. 2009. Indirect reciprocity provides only a narrow margin of efficiency for costly punishment. *Nature* **457:** 79–82.
35. Rozin, P., J. Haidt & K. Fincher. 2009. From oral to moral. *Science* **323:** 1179.
36. Gutierrez, R., R. Giner-Sorolla & M. Vasiljevic. 2012. Just an anger synonym? Moral context influences predictors of disgust word use. *Cogn. Emot.* **26:** 53–64.
37. Gutierrez, R. & R. Giner-Sorolla. 2007. Anger, disgust, and presumption of harm as reactions to taboo-breaking behaviors. *Emotion* **7:** 853–868.
38. Phillips, M.L., A.W. Young, C. Senior, *et al.* 1997. A specific neural substrate for perceiving facial expressions of disgust. *Nature* **389:** 495–498.
39. Phillips, M.L., A.W. Young, S.K. Scott, *et al.* 1998. Neural responses to facial and vocal expressions of fear and disgust. *Proc. Roy. Soc. Lond., Ser. B: Biol. Sci.* **265:** 1809.
40. Sprengelmeyer, R., M. Rausch, U.T. Eysel & H. Przuntek. 1998. Neural structures associated with recognition of facial expressions of basic emotions. *Proc. Roy. Soc. Lond., Ser. B: Biol. Sci.* **265:** 1927.

41. Naidich, T.P., E. Kang, G.M. Fatterpekar, *et al.* 2004. The insula: anatomic study and MR imaging display at 1.5 T. *Am. J. Neuroradiol.* **25:** 222.
42. Varnavas, G.G. & W. Grand. 1999. The insular cortex: morphological and vascular anatomic characteristics. *Neurosurgery* **44:** 127.
43. Augustine, J. 1996. Circuitry and functional aspects of the insular lobe in primates including humans. *Brain Res. Rev.* **22:** 229–244.
44. Mufson, E.J. & M. Mesulam. 1982. Insula of the old world monkey. II: afferent cortical input and comments on the claustrum. *J. Comparat. Neurol.* **212:** 23–37.
45. Mesulam, M. & E. Mufson. 1982. Insula of the old world monkey. III: efferent cortical output and comments on function. *J. Comparat. Neurol.* **212:** 38–52.
46. Mesulam, M. 1982. Insula of the old world monkey I. Architectonics in the insulo-orbito-temporal component of the paralimbic brain. *J. Comparat. Neurol.* **212:** 1–22.
47. Allman, J.M., N.A. Tetreault, A.Y. Hakeem, *et al.* 2010. The von Economo neurons in frontoinsular and anterior cingulate cortex in great apes and humans. *Brain Struct. Funct.* **214:** 495–517.
48. von Economo, C. & G. Koskinas. 1925. *Die Cytoarchitectonic der Hirnrinde des Erwachsenen Menschen.* Springer. Berlin.
49. Seeley, W.W., D.A. Carlin, J.M. Allman, *et al.* 2006. Early frontotemporal dementia targets neurons unique to apes and humans. *Ann. Neurol.* **60:** 660–667.
50. Penfield, W. & M.E. Faulk. 1955. The insula. *Brain* **78:** 445.
51. Craig, A. 2009. How do you feel-now? The anterior insula and human awareness. *Nat. Rev. Neurosci.* **10:** 59–70.
52. Ekman, P. & W. Friesen. 1978. *Facial Action Coding System: A Technique for the Measurement of Facial Movement.* Consulting Psychologists Press. Palo Alto, CA.
53. Izard, C. 1971. *The Face of Emotion.* Appleton-Century-Crofts. East Norwalk. USA.
54. Susskind, J., D.H. Lee, A. Cusi, *et al.* 2008. Expressing fear enhances sensory acquisition. *Nat. Neurosci.* **11:** 843–850.
55. Angyal, A. 1941. Disgust and related aversions. *J. Abnormal Social Psychol.* **36:** 393–412.
56. Ekman, P., E. Sorenson & W. Friesen. 1969. Pan-cultural elements in facial displays of emotion. *Science* **164:** 86–88.
57. Izard, C.E. 1994. Innate and universal facial expressions: evidence from developmental and cross-cultural research. *Psychol. Bull.* **115:** 288–299.
58. Chapman, H.A., D.A. Kim, J.M. Susskind & A.K. Anderson. 2009. In bad taste: evidence for the oral origins of moral disgust. *Science* **323:** 1222–1226.
59. Rozin, P., L. Lowery & R. Ebert. 1994. Varieties of disgust faces and the structure of disgust. *J. Personal. Soc. Psychol.* **66:** 870–881.
60. von dem Hagen, E.A.H., J.D. Beaver, M.P. Ewbank, *et al.* 2009. Leaving a bad taste in your mouth but not in my insula. *Soc. Cogn. Affect. Neurosci.* **4:** 379–386.
61. Anderson, A.K., K. Christoff, D. Panitz, *et al.* 2003. Neural correlates of the automatic processing of threat facial signals. *J. Neurosci.* **23:** 5627–5633.
62. Winston, J.S., J. O'Doherty & R.J. Dolan. 2003. Common and distinct neural responses during direct and incidental processing of multiple facial emotions. *NeuroImage* **20:** 84–97.
63. Krolak Salmon, P., M.A. Henaff, J. Isnard, *et al.* 2003. An attention modulated response to disgust in human ventral anterior insula. *Ann. Neurol.* **53:** 446–453.
64. Fusar-Poli, P., A. Placentino, F. Carletti, *et al.* 2009. Functional atlas of emotional faces processing: a voxel-based meta-analysis of 105 functional magnetic resonance imaging studies. *J. Psychiatr. Neurosci.* **34:** 418.
65. Wicker, B., C. Keysers, J. Plailly, *et al.* 2003. Both of us disgusted in my insula: the common neural basis of seeing and feeling disgust. *Neuron* **40:** 655–664.
66. Jabbi, M., J. Bastiaansen & C. Keysers. 2008. A common anterior insula representation of disgust observation, experience and imagination shows divergent functional connectivity pathways. *PLoS ONE* **3:** e2939.
67. Niedenthal, P.M., L.W. Barsalou, P. Winkielman, *et al.* 2005. Embodiment in attitudes, social perception, and emotion. *Personal. Soc. Psychol. Rev.* **9:** 184–211.
68. Niedenthal, P.M. 2008. Emotion concepts. In *Handbook of Emotion.* M. Lewis, J.M. Haviland-Jones & L.F. Barrett, Eds.: 587–600. Guilford. New York.
69. Caruana, F., A. Jezzini, B. Sbriscia-Fioretti, G. Rizzolatti & V. Gallese. 2010. Emotional and social behaviors elicited by electrical stimulation of the insula in the macaque monkey. *Curr. Biol.* **21:** 195–199.
70. Sprengelmeyer, R., A.W. Young, A.J. Calder, *et al.* 1996. Loss of disgust. *Brain* **119:** 1647–1665.
71. Gray, J.M., A.W. Young, W.A. Barker, A. Curtis & D. Gibson. 1997. Impaired recognition of disgust in Huntington's disease gene carriers. *Brain* **120:** 2029–2038.
72. Sprengelmeyer, R., A. Young, I. Pundt, *et al.* 1997. Disgust implicated in obsessive-compulsive disorder. *Proc. R. Soc. B: Biol. Sci.* **264:** 1767–1773.
73. Corcoran, K., S. Woody & D. Tolin. 2008. Recognition of facial expressions in obsessive-compulsive disorder. *J. Anxiety Disorders* **22:** 56–66.
74. Sprengelmeyer, R., U. Schroeder, A.W. Young & J.T. Epplen. 2006. Disgust in pre-clinical Huntington's disease: a longitudinal study. *Neuropsychologia* **44:** 518–533.
75. Parker, H., R. McNally, K. Nakayama & S. Wilhelm. 2004. No disgust recognition deficit in obsessive-compulsive disorder. *J. Behav. Ther. Exp. Psychiatr.* **35:** 183–192.
76. Johnson, S., J. Stout, A. Solomon, *et al.* 2007. Beyond disgust: impaired recognition of negative emotions prior to diagnosis in Huntington's disease. *Brain* **130:** 1732–1744.
77. Snowden, J., N. Austin, S. Sembi, *et al.* 2008. Emotion recognition in Huntington's disease and frontotemporal dementia. *Neuropsychologia* **46:** 2638–2649.
78. Milders, M., J.R. Crawford, A. Lamb & S.A. Simpson. 2003. Differential deficits in expression recognition in gene-carriers and patients with Huntington's disease. *Neuropsychologia* **41:** 1484–1492.
79. Calder, A., J. Keane, F. Manes, N. Antoun & A. Young. 2000. Impaired recognition and experience of disgust following brain injury. *Nat. Neurosci.* **3:** 1077–1078.
80. Straube, T., A. Weisbrod, S. Schmidt, *et al.* 2010. No impairment of recognition and experience of disgust in a patient

with a right-hemispheric lesion of the insula and basal ganglia. *Neuropsychologia* **48:** 1735–1741.
81. Beckstead, R.M. & R. Norgren. 1979. An autoradiographic examination of the central distribution of the trigeminal, facial, glossopharyngeal, and vagal nerves in the monkey. *J. Comparat. Neurol.* **184:** 455–472.
82. Beckstead, R.M., J.R. Morse & R. Norgren. 1980. The nucleus of the solitary tract in the monkey: projections to the thalamus and brain stem nuclei. *J. Comparat. Neurol.* **190:** 259–282.
83. Mufson, E.J. & M. Mesulam. 1984. Thalamic connections of the insula in the rhesus monkey and comments on the paralimbic connectivity of the medial pulvinar nucleus. *J. Comparat. Neurol.* **227:** 109–120.
84. Petrides, M. & D.N. Pandya. 1994. Comparative architectonic analysis of the human and the macaque frontal cortex. In *Handbook of Neuropsychology*, Vol. 9. F. Boller & J. Grafman, Eds.: 17–58. Elsevier. New York.
85. Pritchard, T.C., R.B. Hamilton, J.R. Morse & R. Norgren. 1986. Projections of thalamic gustatory and lingual areas in the monkey, macaca fascicularis. *J. Comparat. Neurol.* **244:** 213–228.
86. Small, D.M. 2010. Taste representation in the human insula. *Brain Struct. Funct.* **214:** 551–561.
87. Smith-Swintosky, V.L., C.R. Plata-Salaman & T.R. Scott. 1991. Gustatory neural coding in the monkey cortex: stimulus quality. *J. Neurophysiol.* **66:** 1156–1165.
88. Verhagen, J.V. & L. Engelen. 2006. The neurocognitive bases of human multimodal food perception: sensory integration. *Neurosci. Biobehav. Rev.* **30:** 613–650.
89. Kadohisa, M., E.T. Rolls & J.V. Verhagen. 2005. Neuronal representations of stimuli in the mouth: the primate insular taste cortex, orbitofrontal cortex and amygdala. *Chem. Sens.* **30:** 401–419.
90. Small, D., M. Gregory, Y. Mak, *et al.* 2003. Dissociation of neural representation of intensity and affective valuation in human gustation. *Neuron* **39:** 701–711.
91. Scott, T.R. & C.R. Plata-Salam·n. 1999. Taste in the monkey cortex. *Physiol. Behav.* **67:** 489–511.
92. Yaxley, S., E.T. Rolls & Z.J. Sienkiewicz. 1990. Gustatory responses of single neurons in the insula of the macaque monkey. *J. Neurophysiol.* **63:** 689.
93. Verhagen, J.V., M. Kadohisa & E.T. Rolls. 2004. Primate insular/opercular taste cortex: neuronal representations of the viscosity, fat texture, grittiness, temperature, and taste of foods. *J. Neurophysiol.* **92:** 1685.
94. Schienle, A., R. Stark, B. Walter, *et al.* 2002. The insula is not specifically involved in disgust processing: An FMRI study. *Neuroreport* **13:** 2023–2026.
95. Wright, P., G. He, N. Shapira, *et al.* 2004. Disgust and the insula: FMRI responses to pictures of mutilation and contamination. *Neuroreport* **15:** 2347–2351.
96. Stark, R., A. Schienle, M. Sarlo, *et al.* 2005. Influences of disgust sensitivity on hemodynamic responses towards a disgust-inducing film clip. *Int. J. Psychophysiol.* **57:** 61–67.
97. Benuzzi, F., F. Lui, D. Duzzi, *et al.* 2008. Does it look painful or disgusting? Ask your parietal and cingulate cortex. *J. Neurosci.* **28:** 923–931.
98. Fitzgerald, D.A., S. Posse, G.J. Moore, *et al.* 2004. Neural correlates of internally-generated disgust via autobiographical recall: a functional magnetic resonance imaging investigation. *Neurosci. Lett.* **370:** 91–96.
99. Moll, J., R. de Oliveira-Souza, F. Moll, *et al.* 2005. The moral affiliations of disgust: a functional MRI study. *Cogn. Behav. Neurol.* **18:** 68–78.
100. Schaich Borg, J., D. Lieberman & K.A. Kiehl. 2008. Infection, incest, and iniquity: investigating the neural correlates of disgust and morality. *J. Cogn. Neurosci.* **20:** 1529–1546.
101. Stark, R., M. Zimmermann, S. Kagerer, *et al.* 2007. Hemodynamic brain correlates of disgust and fear ratings. *NeuroImage* **37:** 663–673.
102. Schaefer, A., V. Leutgeb, G. Reishofer, *et al.* 2009. Propensity and sensitivity measures of fear and disgust are differentially related to emotion-specific brain activation. *Neurosci. Lett.* **465:** 262–266.
103. Mataix-Cols, D., S.K. An, N.S. Lawrence, *et al.* 2008. Individual differences in disgust sensitivity modulate neural responses to aversive/disgusting stimuli. *Eur. J. Neurosci.* **27:** 3050–3058.
104. Stark, R., A. Schienle, B. Walter, *et al.* 2003. Hemodynamic responses to fear and disgust-inducing pictures: an FMRI study. *Int. J. Psychophysiol.* **50:** 225–234.
105. Schienle, A., A. Schafer, R. Stark, B. Walter & D. Vaitl. 2005. Relationship between disgust sensitivity, trait anxiety and brain activity during disgust induction. *Neuropsychobiology* **51:** 86–92.
106. Vytal, K. & S. Hamann. 2010. Neuroimaging support for discrete neural correlates of basic emotions: a voxel-based meta-analysis. *J. Cogn. Neurosci.* **22:** 2864–2885.
107. Schienle, A., A. Schafer, A. Hermann, *et al.* 2006. FMRI responses to pictures of mutilation and contamination. *Neurosci. Lett.* **393:** 174–178.
108. Stark, R., A. Schienle, C. Girod, *et al.* 2005. Erotic and disgust-inducing pictures-differences in the hemodynamic responses of the brain. *Biol. Psychol.* **70:** 19–29.
109. Schienle, A., A. Schafer, R. Stark, *et al.* 2005. Neural responses of OCD patients towards disorder-relevant, generally disgust-inducing and fear-inducing pictures. *Int. J. Psychophysiol.* **57:** 69–77.
110. Helzer, E.G. & D.A. Pizarro. 2011. Dirty liberals! Reminders of physical cleanliness influence moral and political attitudes. *Psychol. Sci.* **22:** 517–522.
111. Inbar, Y., D.A. Pizarro, J. Knobe & P. Bloom. 2009. Disgust sensitivity predicts intuitive disapproval of gays. *Emotion* **9:** 435.
112. Critchley, H.D., S. Wiens, P. Rotshtein, *et al.* 2004. Neural systems supporting interoceptive awareness. *Nat. Neurosci.* **7:** 189–195.
113. Mehnert, U., S. Boy, J. Svensson, *et al.* 2008. Brain activation in response to bladder filling and simultaneous stimulation of the dorsal clitoral nerve–an FMRI study in healthy women. *NeuroImage* **41:** 682–689.
114. Vandenbergh, J., P. DuPont, B. Fischler, *et al.* 2005. Regional brain activation during proximal stomach distention in humans: a positron emission tomography study. *Gastroenterology* **128:** 564–573.

115. Arnow, B.A., J.E. Desmond, L.L. Banner, *et al.* 2002. Brain activation and sexual arousal in healthy, heterosexual males. *Brain* **125:** 1014–1023.
116. Herde, L., C. Forster, M. Strupf & H.O. Handwerker. 2007. Itch induced by a novel method leads to limbic deactivations: Functional MRI study. *J. Neurophysiol.* **98:** 2347–2356.
117. Weissman, D.H., K.C. Roberts, K.M. Visscher & M.G. Woldorff. 2006. The neural bases of momentary lapses in attention. *Nat. Neurosci.* **9:** 971–978.
118. Cole, M.W. & W. Schneider. 2007. The cognitive control network: integrated cortical regions with dissociable functions. *NeuroImage* **37:** 343–360.
119. Huettel, S.A., C.J. Stowe, E.M. Gordon, *et al.* 2006. Neural signatures of economic preferences for risk and ambiguity. *Neuron* **49:** 765–775.
120. Paulus, M.P., C. Rogalsky, A. Simmons, *et al.* 2003. Increased activation in the right insula during risk-taking decision making is related to harm avoidance and neuroticism. *NeuroImage* **19:** 1439–1448.
121. Binder, J., E. Liebenthal, E. Possing, *et al.* 2004. Neural correlates of sensory and decision processes in auditory object identification. *Nat. Neurosci.* **7:** 295–301.
122. Farb, N.A.S., Z.V. Segal & A.K. Anderson. Submitted. Attention-related modulation of primary interoceptive and exteroceptive cortices.
123. Craig, A. 2003. Interoception: the sense of the physiological condition of the body. *Current Opinion in Neurobiology* **13:** 500–505.
124. Bloom, P. 2004. *Descartes' Baby: How the Science of Child Development Explains What Makes us Human.* Basic Books. New York.
125. Danovitch, J. & P. Bloom. 2009. Children's extension of disgust to physical and moral events. *Emotion* **9:** 107–112.
126. Cannon, P.R., S. Schnall & M. White. 2011. Transgressions and expressions: affective facial muscle activity predicts moral judgments. *Soc. Psychol. Personal. Sci.* **2:** 325–331.
127. Jones, A. & J. Fitness. 2008. Moral hypervigilance: the influence of disgust sensitivity in the moral domain. *Emotion* **8:** 613–627.
128. Eskine, K.J., N.A. Kacinik & J.J. Prinz. 2011. A bad taste in the mouth. *Psychol. Sci.* **22:** 295–299.
129. Wheatley, T. & J. Haidt. 2005. Hypnotic disgust makes moral judgments more severe. *Psychol. Sci.* **16:** 780–784.
130. Schnall, S., J. Haidt, G.L. Clore & A.H. Jordan. 2008. Disgust as embodied moral judgment. *Personal Soc. Psychol. Bull.* **34:** 1096–1109.
131. Royzman, E. & R. Kurzban. 2011. Minding the metaphor: the elusive character of moral disgust. *Emot. Rev.* **3:** 269.
132. Young, L. & R. Saxe. 2011. When ignorance is no excuse: different roles for intent across moral domains. *Cognition* **120:** 202–214.
133. Horberg, E.J., C. Oveis, D. Keltner & A.B. Cohen. 2009. Disgust and the moralization of purity. *J. Personal. Soc. Psychol.* **97:** 963–976.
134. Nabi, R. 2002. The theoretical versus the lay meaning of disgust: implications for emotion research. *Cogn. Emot.* **16:** 695–703.
135. Greene, J.D., L. Nystrom, A. Engell, *et al.* 2004. The neural bases of cognitive conflict and control in moral judgment. *Neuron* **44:** 389–400.
136. Poldrack, R.A. 2006. Can cognitive processes be inferred from neuroimaging data? *Trend. Cogn. Sci.* **10:** 59–63.
137. Parkinson, C., W. Sinnott-Armstrong, P.E. Koralus, *et al.* 2011. Is morality unified? Evidence that distinct neural systems underlie moral judgments of harm, dishonesty, and disgust. *J. Cogn. Neurosci.* **23:** 3162–3180.
138. Schaich Borg, J., W. Sinnott-Armstrong, V.D. Calhoun & K.A. Kiehl. 2011. Neural basis of moral verdict and moral deliberation. *Soc. Neurosci.* **6:** 1–16.
139. Mitchell, J.P. 2009. Inferences about mental states. *Philos. Trans. R. Soc. Lond.: Ser. B, Biol. Sci.* **364:** 1309–1316.
140. Saxe, R. & N. Kanwisher. 2003. People thinking about thinking people. The role of the temporo-parietal junction in "Theory of Mind". *NeuroImage* **19:** 1835–1842.
141. Young, L., F. Cushman, M. Hauser & R. Saxe. 2007. The neural basis of the interaction between theory of mind and moral judgment. *Proc. Natl. Acad. Sci. U.S.A.* **104:** 8235–8240.
142. Davidson, R.J., P. Ekman, C.D. Saron, *et al.* 1990. Approach-withdrawal and cerebral asymmetry: emotional expression and brain physiology. I. *J. Personal. Soc. Psychol.* **58:** 330–341.
143. Berle, D. & E.S. Phillips. 2006. Disgust and obsessive-compulsive disorder: an update. *Psychiatry* **69:** 228–238.
144. Charash, M., D. McKay & N. Dipaolo. 2006. Implicit attention bias for disgust. *Anxiety* **19:** 353–364.
145. Buodo, G., M. Sarlo & D. Palomba. 2002. Attentional resources measured by reaction times highlight differences within pleasant and unpleasant, high arousing stimuli. *Motivat. Emot.* **26:** 123–138.
146. Cisler, J., B. Olatunji, J.M. Lohr & N.L. Williams. 2008. Attentional bias differences between fear and disgust: implications for the role of disgust in disgust-related anxiety disorders. *Cognit. Emot.* **23:** 675–687.
147. Liu, L. & S.A. Simon. 1998. Responses of cultured rat trigeminal ganglion neurons to bitter tastants. *Chem. Sens.* **23:** 125–130.
148. Lim, J. & B.G. Green. 2007. The psychophysical relationship between bitter taste and burning sensation: evidence of qualitative similarity. *Chem. Sens.* **32:** 31–39.
149. Chapman, H.A., D. Lee, J. Susskind, *et al.* The face of distaste. In review.

Ann. N.Y. Acad. Sci. ISSN 0077-8923

ANNALS OF THE NEW YORK ACADEMY OF SCIENCES
Issue: *The Year in Cognitive Neuroscience*

# Color consilience: color through the lens of art practice, history, philosophy, and neuroscience

Bevil R. Conway

Neuroscience Program, Wellesley College, Wellesley, Massachusetts and Department of Neurobiology, Harvard Medical School, Boston, Massachusetts

Address for correspondence: Dr. Bevil R. Conway, Wellesley College – Neuroscience, 106 Central St., Wellesley, MA 02481. bconway@wellesley.edu

**Paintings can be interpreted as the product of the complex neural machinery that translates physical light signals into behavior, experience, and emotion. The brain mechanisms responsible for vision and perception have been sculpted during evolution and further modified by cultural exposure and development. By closely examining artists' paintings and practices, we can discover hints to how the brain works, and achieve insight into the discoveries and inventions of artists and their impact on culture. Here, I focus on an integral aspect of color, color contrast, which poses a challenge for artists: a mark situated on an otherwise blank canvas will appear a different color in the context of the finished painting. How do artists account for this change in color during the production of a painting? In the broader context of neural and philosophical considerations of color, I discuss the practices of three modern masters, Henri Matisse, Paul Cézanne, and Claude Monet, and suggest that the strategies they developed not only capitalized on the neural mechanisms of color, but also influenced the trajectory of western art history.**

**Keywords:** cones; retina; V4; macaque monkey; Modernism; Matisse; Cézanne; Monet

*It is only after years of preparation that the young artist should touch color—not color used descriptively, that is, but as a means of personal expression.*[1]
*A great modern attainment is to have found the secret of expression by color.*[2]

HENRI MATISSE

## Introduction

For Henri Matisse, painting was serious business. Naturally, he wore a suit to work. Matisse, an icon of modern art who "everyone agrees deserves the title of the century's greatest colorist,"[3] lived in the south of France and often painted with a smock to protect his formal attire from sticky oil paint. Wearing a smock was an easy decision. But what about the decisions that followed? As Matisse so bluntly states, "Anyone who paints has to make choices minute by minute."[4] Faced with a blank canvas, how did Matisse decide what marks to apply where and with what color? And how did these decisions go on to shape the trajectory of art history? To some extent, the kind of marks one makes is inevitably determined by body mechanics: our arms are attached at a fixed point, the shoulder, and so any attempt at a straight line invariably results in a gentle curve.[5] The physical structure of our bodies therefore influences drawing practice—and these influences may extend to cognitive development, shaping how we think. Similarly, the way in which our nervous systems encode light signals necessarily determines what we see and how artists paint. An emerging field of research, vision and art, explores the interface between the neural mechanisms of vision and art.[6–10] An extension of this field, which I take up here, concerns the interaction between visual processing and art practice: how do the mechanisms of vision influence the decisions of the artist at work? And what do the strategies that artists employ in making their work tell us about brain function? While the work and comments of artists are not scientific documents, with appropriate sensitivity to their limitations, these materials may be useful in informing

doi: 10.1111/j.1749-6632.2012.06470.x
 © 2012 New York Academy of Sciences. 

our understanding of how the brain works. In this essay, I take up the relationship between painting and color vision.

People have been experimenting with paints and pigments for the entire history of human culture, and have therefore generated a lot of data. But the richest data has come relatively recently, following on the heels of the industrial revolution, which fueled the development of synthetic dyes and pigments and resulted not only in the wide availability of inexpensive pigments in the latter half of the 19th century but also in an expansion of the color gamut available to the artist[11–14] (see also Ref. 15). With the development of synthetic pigments, the use of particular colors was no longer restricted by the wealth of the artist or sponsor, who previously could flaunt their influence by commissioning pictures containing rare pigments like gold, or better, ultramarine blue made from rare lapis lazuli rock. By 1830, synthetic pigments were widely available. For example, manufactured ultramarine was being churned out by factories across Europe, following the invention in 1826 of an inexpensive method of production by Jean Baptiste Guimet in France and, independently, by Christian Gottlob Gmelin in Germany.[11] In the same way that a technical development, photography,[a] paved the way for one of the most creative periods of painting in art history, the development of synthetic pigments foreshadowed an explosion of possibilities for the use of color. In some sense, the only limitation on the use of color became our neural machinery.

From this perspective, paintings can be interpreted as the product of our brains, specifically as the product of all the complex neural circuits that translate physical light signals into behavior, experience, and emotion. The neural machinery of our brains has been sculpted by many influences, not only during a given individual's development, but also during the history of evolution. By closely examining artists' practices and the paintings they make, scientists and art historians can collaborate to uncover clues to how the brain works and thereby gain insight into how the brain has been influenced by cultural history and, in turn, has shaped that history. Here, I describe some of the clever strategies that artists have used to paint in color, and examine how these strategies exploit and reveal the neural basis for color. In this essay, I will consider three titans of color, Paul Cézanne (1839–1906), Claude Monet (1840–1926), and Henri Matisse (1869–1954). The work of these artists certainly influenced the direction of art history. One theory, to which I return in the last section of this essay, is that the art-historical significance of these artists' work derives from the dynamic interaction between the artist and his work during its production, an interaction that is constrained by neural mechanisms of vision and visual feedback. To fully understand the influence of these artists on art history, one may therefore benefit from knowledge of the neural mechanisms of color, which may themselves be better understood in light of historical and philosophical considerations of color. One might then argue that art practice, art history, neuroscience, and philosophy have undergone a kind of consilience, and are dependent upon each other for a complete account of color.[b]

## Color in the world and in our heads

There are lively philosophical debates about color, concerning whether color is determined by some objective real-world criterion, or rather by the particularities of the viewer.[18] These debates have often boiled down to the uneasy question "do you see red like I see red?" Curiously, we rarely ask whether two people see an object as having the same shape. One standard account for the specialness of color rests on the argument that, unlike an object's shape, it cannot be determined objectively by a physical measuring device like a ruler; the measurements of a penny, for example, reveal it to be round,

[a]Some argued that photography would herald the end of painting, which shows how incompetent we can be at predicting the impact of technology on culture. "From today painting is dead," the French painter Paul Delaroche allegedly concluded in 1839 in response to the development of the daguerreotype photographic technique. Although there is no evidence that Delaroche actually said this, the sentiment has been repeated periodically ever since the middle of the 19th century.[1]

[b]The term "consilience" was popularized by Wilson,[17] and I use it to point to the productive intersection between arts and sciences. The various disciplines might not wholly agree, but it is not necessary that they do. Rather, my argument is that understanding will come through appreciation of the many facets of color revealed by many ways of knowing.

confirming our perception of it as round. But this account is flawed on two counts: first, we actually can measure the physical basis for color almost as easily as we can for shape, by using a spectroradiometer to determine the relative fraction of different wavelengths reflected or emitted from a colored object or source. And second, although we may be able to measure the diameter of a penny, rarely is the two-dimensional retinal projection of the penny round; it is almost always an ellipse. Although the physical basis for both can be measured just as accurately, the relationship between these physical stimuli and the perceptions elicited by them would therefore seem to be as complicated in the case of shape as it is in the case of color. The question, then, is why our perception of the shape of the penny is considered universal ("it is round"), yet its color is up for debate ("do you see it as orange like I see it?"). I would argue that the discrepancy between our account of shape and color reveals something unique about color, but it is not that shape is physically measurable and color is not. The discrepancy suggests, instead, that color carries a qualitatively different behavioral valence than shape: we care more about the troubling relationship between the physical basis for color and our experience of it than we do about the relationship between the physics of object shape and our perception of shape. This special feature of color may rest on the fact that we come to know color only through our eyes and not through our muscles and fingertips, which are also used to ascertain shape. In any event, color's specialness may account for the passionate debates that constitute the history of color vision research and continue to this day.

What is so special about color? Scientific studies have shown that humans (with normal trichromatic color) possess extraordinary color detection and discrimination abilities;[19] some claim that "color is what the eye sees best."[20] So we can conclude that color is an important part of our visual experience.[21] But for what, exactly, is color important? Answers often focus on the relevance of color for object recognition (the "ripe fruit" argument[19,22–24]) and occasionally on the use of color for intraspecific communication (the "your face is red, you must be angry" argument[25,26]). The seemingly unquantifiable, qualitative aspect of color that these hypotheses miss is the fact that we *like* color.[27] Color, unlike other aspects of vision (with the possible exception of pleasant faces), appears to have a direct impact on the limbic system. Although emotional reactions can be elicited from shapes (e.g., the outline of a snake), these associations are learned and do not have the same pop-out characteristic of color. Consider a field of 2s in which one "5" is distributed. To identify this unique character, people will typically use a time-consuming search strategy, interrogating each letter ("is this a 5?"). But if the "5" is red (and the distractor 2s are black), the 5 will pop out instantly. For this reason, color is considered a "low-level" or basic visual feature.

Color contributes directly to emotional state,[28–31] which may account for why sports teams with red uniforms win more often.[32] Moreover, people who lose color perception as a result of brain damage become profoundly depressed[33] (although it is obviously unlikely that impaired color is the root cause of most depression). Evidence in support of the intimate relationship between color and emotion/reward fell out of a study examining experimental deep-brain stimulation (DBS) for the treatment of intractable depression. Mayberg and her colleagues found that DBS in humans of a brain region implicated in depression, Area 25, resulted in elevations in mood and enhancement of color perception. Following DBS of Area 25, "all patients spontaneously reported acute effects including ... sharpening of visual details and intensification of colors in response to electrical stimulation."[34] The special status of color may underlie our use of color as a metaphor for emotion and a host of other ineffable experiences, such as musical timbre (often called "sound color").

Although debates rage on in contemporary philosophy, there are some facts concerning color about which we have consensus. First, with the exception of imaginary colors, color is dependent on the spectral content (wavelength) of light reaching the eye from the outside world. But, as we will see, the relationship between our experience of an object's color and this spectrum is not straightforward. While the color of monochromatic light viewed as a disc surrounded by black ("aperture color") can be predicted,[c] it is difficult, if not impossible, to predict

[c]White light, when split by a prism, is divided into a series of monochromatic wavelengths, each of which can be assigned a color term, from red, for the longest wavelengths, to violet, for the shortest wavelengths (ROYGBIV).

a subject's perception of a given stimulus given an objective physical measurement of the stimulus under natural viewing conditions. You might think that the color of an object could be determined by simply measuring the relative amount of each wavelength reflected from the object, but it cannot. Rather, the color we experience is contingent on the spatial and temporal context in which a stimulus is viewed, and these contextual relationships must be computed by the brain. This leads to a second conclusion of consensus: that our visual systems are implicated in encoding color. The responsible neural mechanisms not only transform light signals into electrical impulses that are the currency of the nervous system, but also generate spatial and temporal comparisons of the light signals across the visual scene and integrate these data with the viewer's expectations, shaped by development, experience, and cultural exposure. In coarse terms, these aspects of the neural machinery have been referred to as "bottom–up" and "top–down"; the former term describes the feed-forward, sequential processing of light signals along the visual pathway from the retina, through the thalamus, and up through various visual cortical regions, and the latter term refers to the influences of cognition and prior experience on the brain's calculation of what color is assigned to the feed-forward signals.[35]

The neat division of processing mechanisms into bottom–up and top–down—a scheme that treats neural signals like batons passed from one runner to the next in a relay with a starting gun and a finish line—is a gross simplification that may turn out to be frankly wrong. Certainly, the spatial metaphor of a "line" of discrete "relays" is incorrect. Increasing evidence shows that brain regions are richly connected by feed-forward and feed-back connections that are engaged seemingly simultaneously,[36–39] so placing any brain region at a discrete "stage" in the processing hierarchy is questionable. A more apt analogy might be one of making soup: visual signals contribute to the perceptual output of the brain just as additional ingredients would shape the flavor of a soup, but their contribution to perception and behavior is influenced by the previous state of the brain and how strong the visual signals are in that context, just as the added flavor of any new ingredient is influenced by what else is in the soup. At the risk of pushing the analogy too far, spices would be the diffuse modulatory inputs, like those that regulate attention and wakefulness, which set the tenor for the whole operation. How bottom–up relays and top–down feedback, or cortical soup making are actually implemented in the brain in the service of color is anything but resolved. But once again there are areas of consensus, and new research using new techniques is shedding light on some stubborn questions, as discussed in the next section.

## Neural mechanisms for color

The retina contains three types of cone photoreceptor cells that are the first steps in the feed-forward computation of color (Fig. 1A). These cells do not encode primary colors, and the brain does not mix the activity of the cones as a painter might mix primary paints. Instead, to encode color, our brains have circuits that compute the relative amount of each type of cone activity across the visual scene. This spatial calculation enables the brain to achieve something called color constancy, the phenomenon that causes our experience of a given object's color to be stable despite changing illumination conditions.[27,40–43] Through color constancy, our brains enable us to see color as part of objects, not contingent on whether we are looking at them under a blue sky or a cloudy sky. These two different viewing conditions would change the physical spectral signals received by the eye, yet our experience of the color of an object does not change that much—fair weather or foul, we consider the apple to be red. For the mathematically inclined, the problem of color constancy can be summarized in straightforward terms:[44] the spectral signals that the eye receives from an object are the product of two variables, the spectral content of the illuminant[d] and the absorptive property of the object. As organisms that attained color vision for some selective advantage, the only thing we really care about is the property

[d]The spectral content of the illuminant describes the relative proportion of wavelengths across the visible spectrum that comprises an illuminant. A spectroradiometer can be used to measure the spectral content of the illuminant reflected off a standard "white" card. Natural daylight on a cloudless day has relatively uniform levels of light across the visible spectrum, and will contain a higher proportion of short-wavelength light than tungsten light, which has relatively low amounts of short-wavelength light and a large proportion of long-wavelength light.

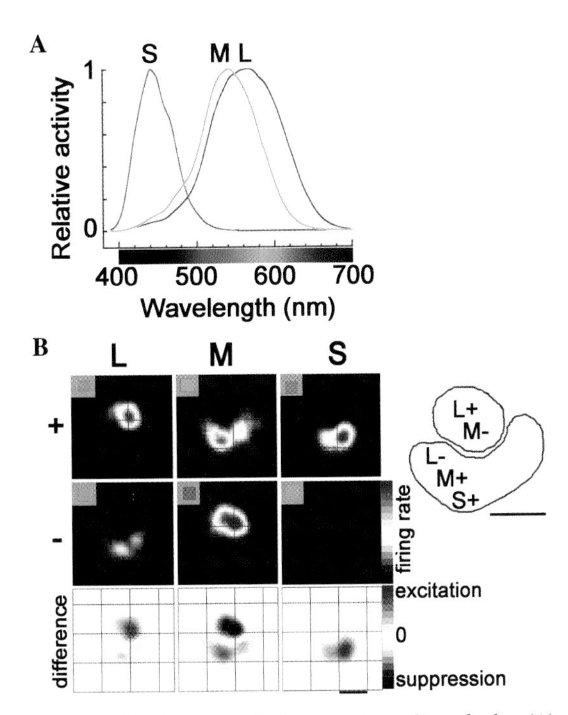


**Figure 1.** The first stages in the neural encoding of color. (A) Cone-absorption spectra of the three classes of cones (L, M, and S) in the retina. (B) Receptive-field of a double-opponent cell in primary visual cortex. Top panels show the spatial receptive-field map generated using sparse noise cone-isolating stimuli and reverse correlation; difference maps show the "+" maps subtracted from the "–" maps. The insets give an indication of the color of each stimulus (although the actual stimuli were presented on a computer monitor and carefully color calibrated). Scale of the grid is 0.75° of visual angle. The receptive-field center was excited by an increase in L cone activity (L+) or a decrease in M activity (M–), and suppressed by a decrease in L (L–) or an increase in M (M+); the receptive-field surround gave the opposite pattern of chromatic tuning. The surround but not the center was modulated by S cones; the S response had the sign as the response to M cones. The diagram to the right provides a summary. Adapted from Refs. 50 and 51.

of the object—the ripe apple needs to be seen to be ripe under all viewing conditions for the experience of color to be evolutionarily advantageous. Because the spectrum of the various lighting conditions can vary enormously, the spectrum coming from the same object under different lighting conditions also varies tremendously. That the brain extracts a more-or-less constant color signal bound to objects despite this changing illumination is a remarkable achievement, one that even the most advanced cameras can only approximate.

The three cone types in the retina are called "L," "M," and "S," because they have peak sensitivities in the long-, middle-, and short-wavelength regions of the visible spectrum. Importantly, however, each cone type has a very broad absorption curve. In the case of the M and L cones, this means that light of virtually every wavelength from the shortest (or bluest) part of the spectrum all the way to the longest (or reddest) part of the spectrum can be effective at eliciting responses. The peak of both the M and L cone types, which historically have been loosely referred to as the "green" and "red" cones, is actually in the yellow part of the spectrum. It is sufficient to say that the rich spectral information hitting the retina is reduced to three numbers at any given retinal location: the amount of activity in the L, M, and S cones. These three signals are compared by retinal bipolar cells just one synapse downstream of the cones, in a process that is thought to involve two channels: one comparing L and M signals and one comparing S signals to the sum of L + M signals. These two channels are still referred to as "red-green" and "blue-yellow," but these short-hand terms are inaccurate because the chromatic tuning of the neurons does not map onto the basic perceptual categories of red, green, blue, and yellow. The optimal chromatic stimulus for the "blue–yellow" channel, for example, is actually lavender–lime. To date, the neural basis for the basic perceptual categories is still unknown,[45,46] although there is some evidence that implicates specialized brain regions downstream of primary visual cortex in the visual processing hierarchy, as I will describe at the end of this section.

The cone-comparison signals encoded by the bipolar cells are converted into a digital signal of action potentials by the retinal ganglion cells, whose axons constitute the optic nerve that courses out the back of the eye and terminates in the lateral geniculate nucleus (LGN) of the thalamus, a structure composed of layers of neural tissue folding, or "genuflecting," to form a peanut-sized structure located deep in the brain—you can almost touch it through the roof of the very back of your throat. Neurons of the LGN send their axons to primary visual cortex, the first cortical stage of visual processing, often called V-1 for short, and paradoxically located at the very back of your brain, as far away from the eyes as any part of the brain can be. Curiously, the LGN receives almost 10 times as many synapses from V-1 as it does from the retina.[47] Each one of these feedback synapses is wimpy in comparison to the retinal feed-forward synapses, but one can already begin to see the limits of the "relay" analogy described earlier: feedback begins before retinal signals even enter the cerebral cortex. The cortex has already started its soup.

Neurons in the LGN with color-coding properties were first described over 50 years ago in the macaque monkey, a creature who has virtually the same cone types and visual system as humans and has become the standard model of human color vision. Although LGN cells carry chromatic information, Hubel and Wiesel showed that the cells have a peculiar response property: they respond well to full-field colored light, but not to small spots of colored light on contrasting backgrounds. For example, a "red-on" cell would respond to full-field red but not to a red spot on a green background.[48] This is puzzling because it is at odds with what we know about color perception, namely that the color of full-field color, a *Ganzfeld*, is not very salient, while a spot of color on a contrasting background pops right off the page (or screen). Evidence for a spatial transformation of the color signals that could mediate color contrast is first found in V-1, manifest by a specialized population of neurons called double-opponent cells.[49,50] Double-opponent cells respond best to color boundaries, say a red region next to a green region. Each V-1 neuron receives inputs from a restricted patch of the retina which corresponds to a portion of the visual field. This small window on the world is called the cell's "receptive-field." An example of a double-opponent receptive-field is shown in Figure 1B. The cell derives its name from the fact that it consists of two kinds of opposition: chromatic and spatial. The chromatic opponency is manifest by opponent responses, excitation or suppression, to L and M cones, whereas the spatial opponency is manifest as opponent responses to the same cone type in spatially offset regions of the receptive-field. The cell shown in Figure 1B was excited most strongly by a reddish spot presented on a blue–green background. These cells encode local cone ratios, and their importance to color rests on the potential for them to mediate color constancy: note that the response of the cell is relatively unperturbed by a full field of light flooding both the receptive-field center and the receptive-field surround—any excitation caused by the stimulus in one receptive-field subregion is counteracted by suppression caused by the stimulus in the adjacent subregion. One can consider the spectral bias of an illuminant to be a full-field stimulus, affecting all parts of a scene; double-opponent cells, which do not react to full-field stimuli, are therefore capable of discounting the spectral bias of the illuminant.

Double-opponent cells represent a small fraction of the total population of neurons in V-1, which initially led investigators to miss them and then to overlook their significance.[51] But importance is not always reflected in numbers: as Richard Gregory pointed out to me, a vivid chromatic signal was carried by a small fraction of the total bandwidth of the analog television signal (analog TV is now obsolete). The rest of the bandwidth of the analog signal was devoted to higher spatial resolution of black-and-white shapes.

Neurons in V-1 besides double-opponent cells can also carry chromatic information (reviewed in Ref. 52), although it is less clear whether this information is used by the brain to encode hue *per se*, or rather used in the service of object recognition and motion perception. One can imagine that there would be a selective advantage for an ability to identify an object boundary formed by colored regions without encoding the colors forming the boundary—this ability is required to defeat camouflage in which the hue of the various regions comprising the object is actually distracting. Many neurons in V-1 show responses to colored boundaries regardless of the colors forming the boundaries, providing a potential neural substrate for this ability. But very few neurons in V-1 show sharp color tuning; that is, V-1 cells do not tend to respond exclusively to one color. Even a given double-opponent cell is not narrowly color-tuned for small spots presented in the center of the receptive-field. Instead the response properties of these cells are determined by the cone-contrast of the stimuli, which does not correlate directly with color perception: the cell shown in Figure 1B shows L versus M + S opponency. A stimulus that increases the activity of the L cones or decreases that of M cones appears red (bright or dark), whereas a stimulus that decreases the activity of S cones appears lime green and one that increases activity of the S cones appears lavender. The crescent-shaped receptive-field subregions of the cell shown in Figure 1B were excited by increases in the activity of M cones and decreases in that of L cones (both of which appear green), but also excited by increases in the activity of S cones (which appears lavender). The take-home message is that while these cells may contribute to color contrast calculations, clearly they are not by themselves encoding hue.

The functional properties and organization of the relatively large region of cortex devoted to vision outside of V-1 are still very dimly understood in comparison to our knowledge of V-1.[53–57] But broad consensus among experts is that V-1 is not the only cortical region involved in processing color: parts of the second visual area (V-2, immediately adjacent to V-1), large islands within the inferior temporal cortex (IT, a large swath of brain buried under the ears and on the bottom of the brain, which contains classically defined areas V-4, PIT, and posterior TEO), and regions of anterior TE (closer to the front of the brain) all contain neurons that respond preferentially to colored stimuli (Fig. 2A). But the first region beyond V-1 in which neurons with narrow hue tuning have been unequivocally described with single-unit recording (the "gold standard") is PIT (Figs. 2B and 2C), and these neurons are densely clustered in millimeter-sized islands of tissue, dubbed "globs."[53] Experimental estimates of the fraction of color-tuned neurons within the globs approach 90%, which represents a remarkable level of specialization.

As a heuristic, one can consider each of the brain regions described here—retina, LGN, V-1, V-2, IT, and TE—to be involved in constructing a distinct aspect of the color percept.[27,52] The three cone types are the basis for trichromacy; retinal ganglion cells that respond in an opponent fashion to activation of different cone classes are the basis for color opponency; double-opponent neurons in V-1 generate local color contrast and are the building blocks for color constancy; glob cells in IT elaborate the perception of hue; and TE integrates color perception in the context of behavior. Certainly, this sketch is grossly simplified and likely inaccurate. We will need to make many more measurements to understand how the activity of neurons in the various stages relate to perception, and there will be much work unpacking how activity within the entire cortical color circuit influences the processing of incoming signals. But we are further ahead than we were 50 years ago when many scientists thought color could be read out directly from the response properties of LGN cells.

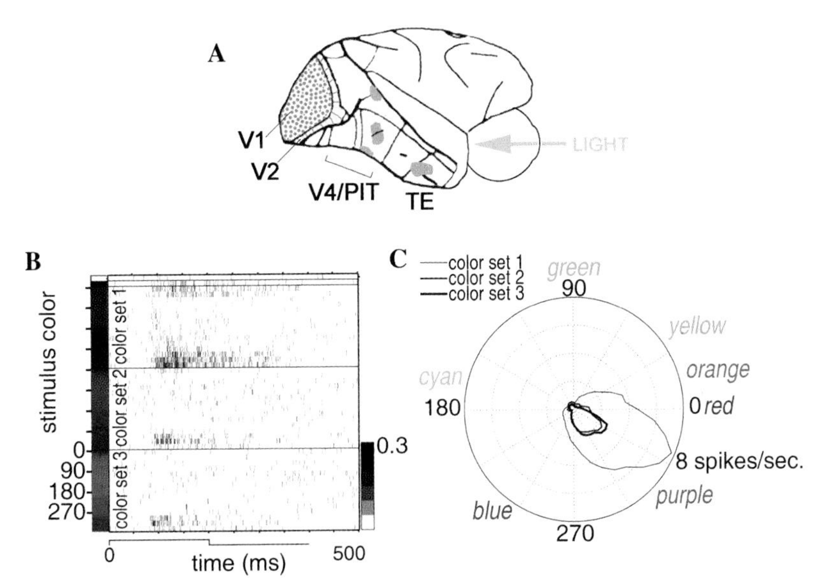


**Figure 2.** (A) Simple hierarchical, feed-forward model of color processing in the macaque cerebral cortex. Regions of cortex shown in gray, which increase in spatial scale along the visual-processing hierarchy from primary visual cortex (V1) to TE, are implicated in color processing. The first region beyond V-1 in which neurons with narrow hue tuning have been unequivocally described with single-unit recording is PIT, where these neurons are densely clustered in millimeter-sized islands of tissue, dubbed "globs." Panels B and C show the color tuning of a typical glob cell. (B) Poststimulus time histograms to an optimally shaped bar of various colors. Responses were determined to white and black (top two rows in histogram) and various colored versions of the bar that varied in brightness. Stimulus onset aligned with 0 ms; stimulus duration (step at bottom). Gray scale bar is average number of spikes per stimulus repeat per bin (1 ms). (C) The color tuning to each of the stimulus sets in polar coordinates. Adapted from Refs. 52 and 53.

## Challenges in painting color

Color poses an enormous challenge for artists because the way a painted mark will ultimately appear is unpredictable. It may be this feature of color that led some artists, like Wassily Kandinsky, to ascribe mystical power to color. It led me to study the neural basis of visual perception in search of explanations. The difficulty of painting in color is juxtaposed with the relative ease with which novice students can acquire representational drawing skills, as Edwards' now classic text attests.[58] The challenge posed by painting in color is exacerbated by the fact that we often conceive of color in simple terms, as a superficial glazing overlaid on an achromatic world. This misconception is perpetuated by children's coloring-in books, which provide black outlines to relay the "important" information, and give deceptively simple instructions to select the appropriate color but to "stay within the lines"; some neuroscientists have characterized the neural encryption of color in these terms.[59]

Curricula at traditional art schools often mirror this process: students are first taught to draw with achromatic pencil lines and then, after mastering black-and-white, to work with paint (as Matisse advised in the epigram quoted at the beginning of this essay). Although this approach may make the problem of representation seemingly tractable, it fosters a misunderstanding of the mechanics of color. The apple in the still life may appear red, but the redness, and our reaction to it, are not attributed solely to the pigment in the skin of the apple, nor can that redness be captured entirely (or even adequately) by matching that pigment with paint. Rather the color is attributed to an unconscious comparison that the visual system makes between the color of the apple, the color of the surrounding regions, and the context in which we see the apple—the very same comparisons that form the basis for color constancy.[e] For whatever reason, our visual system hides this computation from our awareness and leaves us only with an impression that the color is immutable and attached to the object that is the focus of our attention. The consequence is that when confronted with a blank sheet of paper and asked to paint—in color—the apple, we instinctively reach for the tube of red paint, and pay little attention to the background and other viewing conditions. Art teachers find themselves perennially repeating the mantra: "Don't forget about the background!" To which the student retorts: "I'll get to it when I'm done painting the apple." But the application of a background does not have a neutral effect on the work-in-progress, and the surprised and disappointed student is suddenly grossly displeased with the color of the apple, a color that was perfectly satisfactory when floating on the raw white canvas. The experience reinforces the student's resistance to painting backgrounds, and keeps alive the art teacher's mantra.

What accounts for the sudden decline of the painting, coincident with work on the background? The explanation must have something to do with the fact that the student has an impression in his or her mind's eye, a color memory of the object presumably stored deep in brain area TE, which does not match the painting in front of him or her. The dissatisfaction that the student experiences during this exercise underscores how important context is to our experience of color,[f] and how little we acknowledge context in anticipating our experience of color. In the case of color, the spectral bias of the illuminant plays a large role in establishing the

[e]The challenge was articulated by Matisse: "If upon a white canvas I set down some sensations of blue, of green, of red, each new stroke diminishes the importance of the preceding ones. Suppose I have to paint an interior: I have before me a cupboard; it gives me a sensation of vivid red, and I put down a red that satisfies me. A relation is established between this red and the white of the canvas. Let me put a green near the red, and make the floor yellow; and again there will be relationships between the green or yellow and the white of the canvas which satisfy me. But these different tones mutually weaken one another. It is necessary that the diverse marks [signes] I use be balanced so that they do not destroy each other." (see Ref. 2, p. 40).

[f]Color is determined by the context in which it appears, its relationship to the whole, and by the quantity of it within the picture (although this relationship does not bear in a straightforward way on the color). These considerations contribute to the "quantity–quality" calculations described by Bois (1993), and as Bois argues quite compellingly, "The founding principles of Matisse's art proceed from the fact that color relations, which determine expression, are above all relations between surface quantities" (see Ref. 3, p. 23). Thus, Bois argues, Matisse's drawing is determined by the same principles as his color, in service of the same goal of personal expression.

**Figure 3.** Color interaction depends on immediately adjacent colors. The "X" target in panel A has the same spectral reflectance in both panels, yet appears a different color because it is placed on a colored background (adapted from Ref. 75). This effect is demonstrated again in panel C (courtesy of Beau Lotto). Panels B and D show that the color induction effect is obliterated by insulating the target with an achromatic margin.

context for color judgments, and as described earlier, we are virtually blind to the spectral content of the illuminant; this blindness is in very effective service of color constancy. The artist's frustration is therefore a direct consequence of the fact that we evolved to see color, not to paint it. Matisse and other artists indicate cultural progress in this regard, as Matisse clearly acknowledged, "Each color in the painting was determined by and dependent on the others,"[60] and as Bois goes on to reinforce, "Matisse was constantly forced to start from scratch because each color stroke implied a further dissonance, ricochet-like, and necessitated an unsettling of the picture's global color harmony."[61]

The neglect of context is not unique to perceptions of color: the anticipated emotional reaction we have to a given event is radically impacted by the context in which the event is experienced, although we rarely pay much attention to these effects.[62] Color, like a mental preview of a future event, is essentialized, defined by salient aspects and largely agnostic of context. We define color by its category (red, orange, yellow, green, and blue), one that may have behavioral or linguistic relevance. But the context, as the art student's experience shows, is ignored despite the fact that context significantly shapes not only our color experience, but also our emotional reaction. This complexity makes the task of artists particularly difficult, for they must uncover the unconscious mechanisms that underlie color in order to be able to paint in color.[63]

The context dependence of color means that the colored elements comprising the scene interact; and the neural basis for these interactions almost certainly depends to some extent on the calculations made by double-opponent cells. The interaction of color constituted the primary focus of the artist and teacher Josef Albers (1888–1976), whose powerful images underscored the disjunction between what we think underlies a color and the role of context. Figure 3A shows one of Albers' famous color induction demonstrations. In this image, Albers is employing a simple color contrast effect to alter the appearance of a line by placing the line against different colored backgrounds. The image remains surprising even though we fully understand the power of color context to shape our color experience. And all the information in the world is still insufficient to stave off the question, "But what color is the line in reality?"—as if our eyes have deceived us on this rare occasion. Rather than deception, these demonstrations showcase what the visual system is constantly doing—and usually to great adaptive effect—by taking context into account, and then discounting that very information. Such contrast effects continue to inspire the development of powerful demonstrations, like Beau Lotto's cube shown in the Figure 3C. These contextual interactions continue to be mined for clues to the operation of the visual system.[64–66]

Why is it so difficult to learn how to render perceptually accurate color? There are probably several reasons, but I think the main one has to do with the fact that the color computation is effortless. The automatic nature of color leads us to believe that there is a straightforward relationship between the physical stimulus (photons off the apple, and paint in the artist's case) and the color experience. This is reflected in the instructions often given to art students. Indeed, when I took art courses at university, this was the instruction I received: identify a patch of color in the scene, mix it on your palette, and apply it at the appropriate location on the canvas. The instruction sounds logical but it is deeply flawed. One mixes the color on a gray or white palette and applies the paint to a canvas that is white or gray (at least to start) yet the color is ultimately viewed in a colored context. If one were asked to paint Lotto's cube, one might logically begin by painting the yellow squares with paint that appears yellow on the raw white canvas, instead of using paint that appears gray. This false start would influence all subsequent color decisions, and the outcome would consist of profoundly distorted color relationships.

Painting color accurately requires access to information that is simply not available to our perceptual apparatus, which may be why paint-by-number kits are so compelling (Fig. 4). Putting a milky greenish-brown patch in the sky does not seem like the right thing to do at the outset of the project, yet somehow the color is appropriate in the context of the completed image. Working through the painting becomes a kind of joke, in which the punch line (what color the marks appear to have in the finished picture) is surprising.

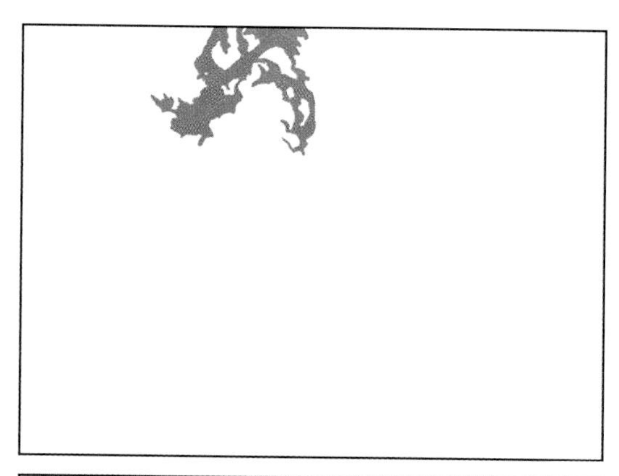

**Figure 4.** Paint-by-number image, partially completed (top panel), and finished (bottom panel).

## Color-master strategies

So how do the painters we think of as masterful colorists capture perceptually accurate color relationships, and what do their strategies tell us about how the nervous system functions? The very category of "colorist" was generated within modern art tending toward abstraction. As painters turned from the strictures of representation that photography had mastered, they were liberated to focus on color, a more recondite aspect of perception. Perhaps we can get some insight into this conjoined project involving both culture and the nervous system from the artists' unfinished works. The left panel of Figure 5 shows a painting that Paul Cézanne was working on at the time of his death. What is striking about it is the lack of a defined subject. Unlike the academic painters of his day, Cézanne does not begin with a well-resolved preparatory line drawing; instead, he begins the work almost immediately with intensely colored paint, and continues to develop the image with patches of color distributed over the surface of the painting. We can begin to make out the suggestion of tree trunks, foliage, sky, and earth—many of the cues to such suggestions are purely coloristic. Using this approach, Cézanne recruits his visual system as a meter of accurate color. Importantly, he has given up the natural conviction that the color of the patch is stable throughout the

**Figure 5.** The visual system as feedback device used to paint color: Paul Cézanne (1839–1906; French) Study of trees 1904 Oil on Canvas; Mont Ste.-Victoire seen from les Lauves [V.798; R.912] 1902–1906.

development of the image. He is able to alter the color through adjacency, by adapting and modifying the painted context in which a colored patch appears. Specifically, he alters the colors of his marks by adding surrounding colors, changing their appearance through changes in context. The approach allows a form of evolution: Cézanne creates a diversity of marks, and then progressively emphasizes those that are the most effective—that most accurately capture his perceptual color experience—and de-emphasizes those that are least effective, by changing their context or occasionally covering a patch with a different color. As the marks on the canvas suggest, this is a dynamic and unpredictable process in which the decision of what colored patch to apply next is largely prescribed by local circumstances—what mark he makes is determined by what is needed to compensate for the change in color appearance of all the marks on the canvas caused by the mark just made. Although the trajectory of the painting may be set (a painting of trees), Cézanne does not seem to be attempting to reproduce a completed picture he has in his mind's eye at the outset of the painting. Rather, the precise combination of colors and marks he uses is made up on the spot and guided continually by visual feedback.

Many colorists have worked in a similarly empirical fashion, developing paintings through trial and error. Perhaps the most famous is Claude Monet, an exemplar well known to Cézanne, who made painting *au plein air* enormously popular in the 19th century. This approach to painting supposedly involves making pictures of scenery outside under natural light, directly looking at the subject being painted. On first blush, this appears to be a similar approach to that used by Cézanne for it would seem to encourage direct visual feedback. So I was surprised to learn that most of Monet's early finished paintings were completed in his studio. (This is, incidentally, a delightful historical example of how the rhetoric produced around art often precedes the art actually made that way.) The painting shown in Figure 6, for example, was begun *au plein air*, but completed some months later in his studio, in winter. Without a real-life scene immediately before him, Monet completed the image through many local decisions, errors, and corrections reflected in the thick layers of paint of the finished picture. Beginning the image *au plein air* honed Monet's observational skills and presumably refined his color memory, and it was this memory that provided stable feedback during the development of the painting in studio. Paradoxically, this feedback might have been more stable than if the scene lay before him in life because real-world scenes are constantly changing, the light shifting throughout the day with a time course shorter than the time required to paint. In his later career, Monet would shift from canvas to canvas throughout the day actually painting *au plein air* before the same motif, making numerous sets of paintings of haystacks and cathedrals that document the passage of time and the changing play of light throughout the day. As a result of his careful observation, and the time spent in studio critically assessing the success of his attempts, Monet was able to reproduce color boundaries as they move through shadow in a way that reflects the cone ratios generated by real color boundaries moving in and out of shadow. It

**Figure 6.** The visual system as feedback device used to paint color: Claude Monet (1840–1926). Femmes au jardin (Women in the Garden), started spring/summer 1866 open air in Ville-d'Avray, finished the following winter in studio Honfleur. Collection Musée D'Orsay, Paris (RF2773) Oil on canvas.

is these local cone ratios that are encoded by the visual system and form the neural building blocks for color. It remains a testable hypothesis whether the cone ratios elicited by the color boundaries in Monet's paintings are similar to those encoded by the nervous system, perhaps by double-opponent cells.

I do not want to suggest that all masters of color work in the same way. Indeed, Monet's early practice is manifestly different than the later methods of Cézanne, although they lived at more or less the same time and painted some of the same motifs in France. In fact, it is the inventiveness of the artist's solution to the problem of color that contributes to the visual interest of their work, and their lessons for neuroscience. Consider the artist with which we began, also recognized for his stunning use of color: Henri Matisse. A striking feature in many of Matisse's mature canvases, beginning in early 1904, is that he leaves portions of the raw canvas untouched. These white regions typically fall at the interface between two differently colored marks. Figure 7 shows two examples dating from 1905 to 1912; the spare use of paint, in contrast to the slathered thick overlapping marks deployed by Monet, was a consistent feature of Matisse's process, one also

**Figure 7.** The use of achromatic borders to separate colored regions in paintings of Henri Matisse. Henri Matisse (1869–1954; French): Intérieur à Collioure/La sieste (Interior at Collioure/The Siesta), Oil on canvas, 23 1/4;" × 28 3/8" Private collection, 1905 (left panel). Poissons rouges et sculpture (Goldfish and Sculpture) spring–summer 1912, oil on canvas 45 3/4" x 39 1/2" Museum of Modern Art, NY (right panel).

exploited by Cézanne.[g] Moreover, most of Matisse's canvases were signed, confirming for the art market that Matisse considered them finished works; the white omissions were clearly intended.

What lay behind Matisse's intention? A clue is provided by some classic psychophysical work reviewed by Brenner and his colleagues:[67] "It is known that chromatic induction is primarily determined by the color of directly adjacent surfaces. . .This is consistent with the idea that information at the borders is critical in determining the perceived color." Chromatic induction is the Yin of the chromatic contrast Yang: it is the phenomenon whereby a target that appears an achromatic gray when placed

[g]The importance of the white unfinished spaces to the color relationships was quite obvious to Cézanne, as demonstrated with his exchange with the art dealer and critic Vollard, as Bois recounts, "What Cézanne said to Vollard concerning the two small spots on the hands in his portrait that weren't yet covered with pigment: 'Maybe tomorrow I will be able to find the exact tone to cover up those spots. Don't you see, Monsieur Vollard, that if I put something there by guesswork, I might have to paint the whole canvas over starting from that point?' " Bois goes on to emphasize that "Matisse was well aware that the apparently 'unfinished' quality of Cézanne's canvasses had an essential function in their construction."[61]

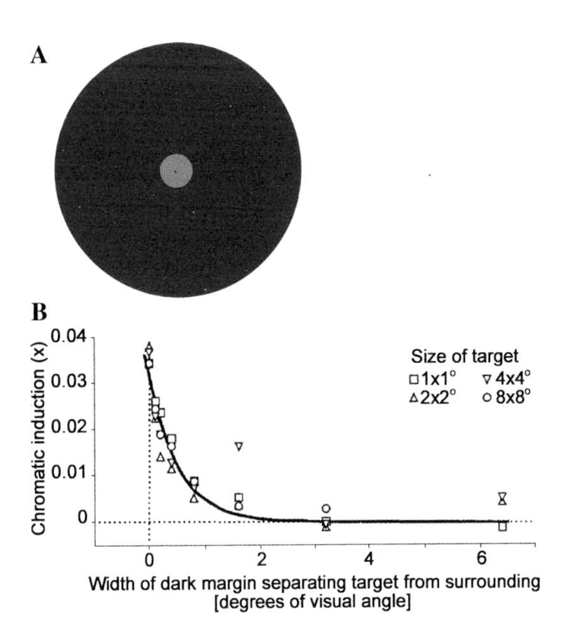


**Figure 8.** (A) Color induction demonstration: see text for details. (B) Color induction is eliminated when achromatic boundaries separate an otherwise achromatic target from the colored surround. The panel shows induction in the $x$ direction (red–green) of the cone space of MacLeod and Boynton as a function of the distance between the border of a target and the colored surround; similar effects are seen in the $y$ direction (blue–yellow). Adapted from Ref. 68.

**Figure 9.** The development of color in the later paintings of Henri Matisse accentuates their inherent flatness. Henri Matisse (1869–1954; French). Left: Small Door of the Old Mill (Petiteporte du viewx moulin), spring-summer 1898, Oil on canvas 18 7/8" x 14 1/8". Private collection. Right: Plum Blossoms, Green Background (La branche de prunier, fond vert), Vence, villa Le Reve [winter-spring], 1948. Oil on canvas 45 5/8" x 35". Private collection. [Correction added after online publication 03/26/2012: the previous sentences were added for clarity.]

on a white (or black) background will appear colored when surrounded by a color field; the induced color is the complementary color of the background. Figure 8A gives an example in which the "gray" central disc appears greenish; curiously, the induced color is often stronger as an afterimage: maintain your gaze on the black dot in the center of the disc, then after 10 seconds transfer your gaze to the black dot to the right. Most observers report the appearance of a striking red spot that is the afterimage to the induced green. Figure 8B, adapted from an original paper by Ref. 68, shows that the degree of color induction (indicated along the *y*-axis) falls off precipitously if the target is separated from the surrounding colored field by an achromatic margin. I have illustrated this observation using the powerful Albers and Lotto demonstrations shown in Figures 3B and D. Given the importance of these local interactions, it is perhaps not surprising that Albers always painted his color fields immediately adjacent to each other. The prediction from Albers is that color is dependent on neural calculations that take place over small local spatial scales, comparable to the spatial scale of double-opponent receptive-fields, which compare the relative cone ratio in one part of visual space with the cone ratio in an immediately adjacent region of visual space.

Let's return to Matisse. Whether he was conscious of it or not, the small achromatic, white gaps left by Matisse would have insulated the appearance of the colors against color induction during the development of the image. As a result, the marks Matisse made remain largely the color he chose them to be at the time he made them. In this way, Matisse was liberated to use vibrant saturated colors, often right out of the tube, without the fear that his colors would become garish from chromatic contrast. The desire to use saturated colors is pervasive, suggesting deep roots in our neural hardware. And the consequence of Matisse's strategy for protecting against induction was, I wager in the next section, an important driving force in the trajectory of art history and the development of Modernism.

## Painting practice, new materials, and Modernism

Beginning with his fauvist work in the early 1900s and extending for the rest of his career, Matisse's output represents a pinnacle of Modernism and owes its success in large part to his innovative use of color.[2,69,70] One consequence of Matisse's dedication to color, in the reading of historians and theorists of modern art, is the accentuation of the inherent flatness of a painted surface. Compare

**Figure 10.** The use of achromatic borders to separate colored regions in paintings of Max Beckmann. Max Beckmann (1884–1950; German): Self portrait in blue jacket 1950, 54 3/4″ x 36″ The Saint Lous Art Museum (right panel). Bequest of Morton D. May. Oil on Canvas Self Portrait in Florence, 1907 (left panel).

two paintings that bookend Matisse's long career (Fig. 9). In the early painting, Matisse seems to struggle to achieve a representation of perspective and depth; the colors he uses are subsidiary. In the later painting, the color is fundamental and more playful, but the image lacks depth, appearing instead as a decorative wall panel not unlike a carpet with broad areas of uniform color enlivened by small graphical marks. Without addressing the cultural considerations as to why Matisse might have sought this flatness, the consensus is that he achieved it. The development of color in this later work, as already described for the images shown in Figure 7, involves a sparing use of paint that leaves bare portions of the underlying canvas, particularly at boundaries between colored regions. As described earlier, this strategy insulates Matisse's color choices against color induction during the process of painting, so the colors we see in the final painting are presumably very close to the colors Matisse decided upon when the picture was being created. Matisse's simple strategy of limiting color induction by leaving white margins around the colored marks was adopted by many other Modernists, including Wassily Kandinsky, Milton Avery, Mark Rothko, and later Frank Stella, whose "pinstripe" paintings launched his career in 1959 with their pulsating whispery white "breathing lines," the remnant unpainted white buffers between broadly painted black stripes (Ref. 71 discussed in Ref. 72). Others, like Max Beckmann, employed a variant: the use of heavy black lines (Fig. 10), which serve the same purpose. Like Matisse, Beckmann began his career attempting to capture space and depth, but concluded by painting images whose color and flatness are central. In these cases, the vivid flatness of the work would seem to be attributed to an emphasis on color because color is for the most part a surface property: the surface of the painting is flat, a fact we come to be aware of through its color.[h] Indeed, the neural mechanisms of color are tied up with the neural mechanisms responsible for encoding surfaces and textures.[73] This attribute of color may, incidentally, account for the philosophical claim that color is tied to objects.

Modernism characterizes a broad movement in thought and culture in Western society that culmi-

[h]Some have argued for a different spatial register to account for the visual experience of colors deployed in intangible screens, like the sky, a smoky sunset, or a cave wall indirectly illuminated by a fire; whether these colors are experienced as surfaces is perhaps debatable.

nated sometime between the end of the 19th century and the middle of the 20th century. It is during this time that the role of subjective experience in art making and interpretation becomes significant. The grand subjects that were often the basis for pictures made during the Renaissance until the French Revolution, such as history, kings, queens, discoveries, and conquests, are rarely the focus of Modernist pictures. (Modernists tended to break with the academy that persisted with those themes.) Instead, we find depictions of domestic life, intimate scenes that underscore the interaction of the artist and his subject. The work showcases the involvement of people with their new-found self-consciousness. These trends have been exhaustively described in the extensive literature on Modernism. Here, I shift focus to question the relationships between the artist, her materials, and her artistic process—which I construe broadly to include the dynamic interaction between the artist and her work during its development, an interaction that is inextricably characterized by the neural mechanisms of vision and visual feedback. These relationships evolved during Modernism and have continued to be a central force in art making today.

The flatness of Matisse's work advanced the Modernist agenda by drawing attention to the material properties of the work: Matisse's work proclaims itself to be objects made of paint and canvas, rather than mere depictions of scenes. As Wright suggests, Matisse "is becoming more interested in paint than in sight."[70] Matisse's focus on color, and on the material properties of paintings, set the stage for high Modernist masters like Mark Rothko and Barnett Newman, whose mature canvases are entirely abstract, consisting simply of large swaths of bold uniform color. Matisse's work is a bridge between one side "the eye of Impressionism or Neo-Impressionism, all eager regard, scanning the shifting surfaces of the world, and on the other, the eye of modernist visuality, absorbed in the formal qualities of a the painting's own surface."[70] I argue here, Matisse's accomplishment was brought about in part because of the particularities of his practice, one aspect of which sought to insulate color from induction. The result of this particular strategy, those vacant bits of canvas, revealed what the paintings were made of; and as a consequence, the making of paintings, rather than their content, took center stage. As Elderfield describes it, the "substance of painting itself came to fulfill the functions that form and structure had fulfilled earlier";[69] see also Ref. 74.

Matisse's paintings are not appreciated as illusions of real space, but rather as objects whose power to move us, much like the power of color itself, is difficult to pin down. The power of Matisse's pictures surely rests on a successful use of color; and this use of color derives its success in part from the process Matisse adopted in making the pictures, a process that was contingent on the particularities of his practices, and their probing relation to the neural mechanisms of color. In the case of the example described here, the decision to leave portions of the canvas unpainted may be attributed to a practical desire to mitigate color induction; but the consequence of this decision pulled back the curtain on the process of art making itself and emphasized the materials of art, thereby underscoring the subjective experience of the artist as art maker, actively engaged in the making of an object.

Matisse's interest in color propelled his rejection of tradition, a tradition that aimed not only to represent objects in space and depth but also to subjugate the materials of painting—brush, paint, and canvas—to the requirements of mimetic representation. Matisse's approach freed painting from this traditional requirement, and is justly celebrated for that liberation. The artist makes the painting self-conscious by underscoring its own making. The viewer cannot escape appreciating that the painting is made of paint and canvas: we have direct evidence of these facts, revealed in many of his paintings by vacant white lines across their surfaces. Thus, it seems that Matisse's sensitivity to color and the innovative process he developed to make his pictures, helped to advance one important aspect of the Modernist agenda: the desire to achieve an accurate representation of the artist's subjective experience of color.

## Acknowledgments

This work was supported by the Radcliffe Institute for Advanced Study (Harvard University) and the National Science Foundation. I am grateful to Caroline Jones, Ann Jones, David Hilbert, Dale Purves, Alexander Rehding, and an anonymous reviewer for helpful discussions and comments on the manuscript. I thank Rosa Lafer-Sousa for help in preparing the manuscript.

## Conflicts of interest

The author declares no conflicts of interest.

## References


1. Wolff, T.F. Drawings of Matisse. *N.Y. Christian Science Monitor*, 25 March 1985, p. 27.
2. Flam, J. 1995. *Matisse on Art (Revised Edition)*. University of California Press. Berkeley and Los Angeles, CA: p. 121.
3. Bois, Y.-A. 1993. Matisse and "Arche Drawing." In *Painting as Model*. MIT Press. Cambridge, MA: p. 61.
4. Flam, J. 1995. *Matisse on Art (Revised Edition)*. University of California Press. Berkeley and Los Angeles, CA: p. 101.
5. Van Sommers, P. 1984. *Drawing and Cognition: Descriptive and Experimental Studies of Graphic Production Processes*. Cambridge University Press. New York.
6. Cavanagh, P. 2005. The artist as neuroscientist. *Nature* **434:** 301–307.
7. Conway, B.R. & M.S. Livingstone. 2007. Perspectives on science and art. *Curr. Opin. Neurobiol.* **17:** 476–482.
8. Livingstone, M.S. 2002. *Vision and Art: The Biology of Seeing*. Abrams Press. New York.
9. Zeki, S. 1999. *Inner Vision: An Exploration of Art and the Brain*. Oxford University Press. New York.
10. Marmor, M.F. & J.G. Ravin. 2009. *The Artist's Eyes: Vision and the History of Art*. Abrams Books. New York.
11. Roy, A., Ed. 1993. *Artist's Pigments: A Handbook of Their History and Characteristics, Volume 2*. National Gallery of Art. Washington, DC.
12. Feller, R.L., Ed. 1986. *Artist's Pigments: A Handbook of Their History and Characteristics, Volume 1*. National Gallery of Art. Washington, DC.
13. Fitzhugh, E.W., Ed. 1997. *Artists Pigments: A Handbook of Their History and Characteristics, Volume 3*. National Gallery of Art. Washington, DC.
14. Berrie, B.H., Ed. 2007. *Artists' Pigments: A Handbook of Their History and Characteristics, Volume 4*. National Gallery of Art Washington, DC.
15. Douma, M. 2011. Pigments through the ages. URL http://www.webexhibits.org/pigments [accessed on 15 August 2011].
16. Crimp, D. 1981. The end of painting. *October* **16:** 69–86.
17. Wilson, E.O. 1998. *Consilience: The Unity of Knowledge*. Knopf. New York.
18. Byrne, A. & D.R. Hilbert. 1997. Colors and Reflectances. In *Readings on Color Volume 1: The Philosophy of Color*. MIT Press. Cambridge, MA: pp. 263–288.
19. Gegenfurtner, K.R. & J. Rieger. 2000. Sensory and cognitive contributions of color to the recognition of natural scenes. *Curr. Biol.* **10:** 805–808.
20. Chaparro, A., *et al.* 1993. Colour is what the eye sees best. *Nature* **361:** 348–350.
21. Hurlbert, A. 1997. Colour vision. *Curr. Biol.* **7:** R400–402.
22. Dominy, N.J. & P.W. Lucas. 2001. Ecological importance of trichromatic vision to primates. *Nature* **410:** 363–366.
23. Smith, A.C., *et al.* 2003. The effect of colour vision status on the detection and selection of fruits by tamarins (*Saguinus* spp.). *J. Exp. Biol.* **206:** 3159–3165.
24. Caine, N.G. & N.I. Mundy. 2000. Demonstration of a foraging advantage for trichromatic marmosets (*Callithrix geoffroyi*) dependent on food colour. *Proc. R. Soc. Lond. B* **267:** 439–444.
25. Changizi, M.A., Q. Zhang & S. Shimojo. 2006. Bare skin, blood and the evolution of primate colour vision. *Biol. Lett.* **2:** 217–221.
26. Fernandez, A.A. & M.R. Morris. 2007. Sexual selection and trichromatic color vision in primates: statistical support for the preexisting-bias hypothesis. *Am. Nat.* **170:** 10–20.
27. Conway, B.R. 2009. Color vision, cones, and color-coding in the cortex. *Neuroscientist* **15:** 274–290.
28. Valdez, P. & A. Mehrabian. 1994. Effects of color on emotions. *J. Exp. Psychol. Gen.* **123:** 394–409.
29. Adams, F. & C. Osgood. 1973. Cross-cultural study of affective meanings of color. *J. Cross. Cult. Psychol.* **4:** 135–156.
30. Ou, L. *et al.* 2004. A study of colour emotion and colour preference. Part I: Colour emotions for single colours. *Color Res. Appl.* **29:** 232–240.
31. Gao, X., *et al.* 2007. Analysis of cross-cultural color emotion. *Color Res. Appl.* **32:** 223–229.
32. Attrill, M.J., *et al.* 2008. Red shirt colour is associated with long-term team success in English football. *J. Sports Sci.* **26:** 577–582.
33. Sacks, O. & R. Wasserman. 1987. The case of the colorblind painter. *N.Y. Rev. Books* **34:** 25–34.
34. Mayberg, H.S., *et al.* 2005. Deep brain stimulation for treatment-resistant depression. *Neuron* **45:** 651–660.
35. Gregory, R.L. 1998. *Eye and Brain: The Psychology of Seeing*. Oxford University Press. Oxford.
36. Lamme, V.A., H. Super & H. Spekreijse. 1998. Feedforward, horizontal, and feedback processing in the visual cortex. *Curr. Opin. Neurobiol.* **8:** 529–535.
37. Hupe, J.M., *et al.* 1998. Cortical feedback improves discrimination between figure and background by V1, V2 and V3 neurons. *Nature* **394:** 784–787.
38. Angelucci, A., *et al.* 2002. Circuits for local and global signal integration in primary visual cortex. *J. Neurosci.* **22:** 8633–8646.
39. Bullier, J., *et al.* 2001. The role of feedback connections in shaping the responses of visual cortical neurons. *Prog. Brain Res.* **134:** 193–204.
40. Land, E.H. 1977. The retinex theory of color vision. *Sci. Am.* **237:** 108–128.
41. Brainard, D.H. & B.A. Wandell. 1986. Analysis of the retinex theory of color vision. *J. Opt. Soc. Am. A* **3:** 1651–1661.
42. Kraft, J.M. & D.H. Brainard. 1999. Mechanisms of color constancy under nearly natural viewing. *Proc. Natl. Acad. Sci. USA* **96:** 307–312.
43. Hansen, T., S. Walter & K.R. Gegenfurtner. 2007. Effects of spatial and temporal context on color categories and color constancy. *J. Vis.* **7:** 1–15.
44. Stockman, A. & D.H. Brainard. 2010. Color vision mechanisms. In *OSA Handbook of Optics*. 3rd ed. M. Bass, Ed. McGraw-Hill. New York.
45. Valberg, A. 2001. Unique hues: an old problem for a new generation. *Vision Res.* **41:** 1645–1657.

46. Webster, M.A., *et al.* 2000. Variations in normal color vision. II: Unique hues. *J. Opt. Soc. Am. A. Opt. Image Sci. Vis.* **17:** 1545–1555.
47. Briggs, F. & W.M. Usrey. 2010. Corticogeniculate feedback and visual processing in the primate. *J. Physiol.* **589:** 33–40.
48. Wiesel, T.N. & D.H. Hubel. 1966. Spatial and chromatic interactions in the lateral geniculate body of the rhesus monkey. *J. Neurophysiol.* **29:** 1115–1156.
49. Michael, C.R. 1978. Color vision mechanisms in monkey striate cortex: dual-opponent cells with concentric receptive fields. *J. Neurophysiol.* **41:** 572–588.
50. Conway, B.R. 2001. Spatial structure of cone inputs to color cells in alert macaque primary visual cortex (V-1). *J. Neurosci.* **21:** 2768–2783.
51. Conway, B.R. & M.S. Livingstone. 2006. Spatial and temporal properties of cone signals in alert macaque primary visual cortex. *J. Neurosci.* **26:** 10826–10846.
52. Conway, B.R., *et al.* 2010. Advances in color science: from retina to behavior. *J. Neurosci.* **30:** 14955–14963.
53. Conway, B.R., S. Moeller & D.Y. Tsao. 2007. Specialized color modules in macaque extrastriate cortex. *Neuron* **56:** 560–573.
54. Conway, B.R. & D.Y. Tsao. 2006. Color architecture in alert macaque cortex revealed by FMRI. *Cereb. Cortex* **16:** 1604–1613.
55. Harada, T., *et al.* 2009. Distribution of colour-selective activity in the monkey inferior temporal cortex revealed by functional magnetic resonance imaging. *Eur. J. Neurosci.* **30:** 1960–1970.
56. Yasuda, M., T. Banno & H. Komatsu. 2010. Color selectivity of neurons in the posterior inferior temporal cortex of the macaque monkey. *Cereb. Cortex* **20:** 1630–1646.
57. Tanigawa, H., H.D. Lu & A.W. Roe. 2010. Functional organization for color and orientation in macaque V4. *Nat. Neurosci.* **13:** 1542–1548.
58. Edwards, B. 1979. *Drawing on the Right Side of the Brain.* Penguin Putnam. New York.
59. Gegenfurtner, K.R. & D.C. Kiper. 2003. Color vision. *Annu. Rev. Neurosci.* **26:** 181–206.
60. Flam, J. 1995. *Matisse on Art (Revised Edition).* University of California Press. Berkeley and Los Angeles, CA: p. 62.
61. Bois, Y.-A. 1993. Matisse and "Arche Drawing." In *Painting as Model.* MIT Press. Cambridge, MA: p. 48.
62. Gilbert, D.T. & T.D. Wilson. 2009. Why the brain talks to itself: sources of error in emotional prediction. *Philos. Trans. R. Soc. Lond. B Biol. Sci.* **364:** 1335–1341.
63. Hurlbert, A. 2007. Colour constancy. *Curr. Biol.* **17:** R906–907.
64. Lotto, R.B. & D. Purves. 2000. An empirical explanation of color contrast. *Proc. Natl. Acad. Sci. USA* **97:** 12834–12839.
65. Shevell, S.K. & F.A. Kingdom. 2008. Color in complex scenes. *Annu. Rev. Psychol.* **59:** 143–166.
66. Purves, D., W.T. Wojtach & R.B. Lotto. 2011. Understanding vision in wholly empirical terms. *Proc. Natl. Acad. Sci. USA* **108**(Suppl 3): 15588–15595.
67. Brenner, E., *et al.* 2003. Chromatic induction and the layout of colours within a complex scene. *Vision Res.* **43:** 1413–1421.
68. Brenner, E. & F.W. Cornelissen. 1991. Spatial interactions in color vision depend on distances between boundaries. *Naturwissenschaften* **78:** 70–73.
69. Elderfield, J. 1979. Describing Matisse. In *Henri Matisse: A Retrospective.* J. Elderfield, Ed.: 14. Museum of Modern Art. New York.
70. Wright, A. 2004. *Matisse and the Subject of Modernism.* Princeton University Press. Princeton and Oxford.
71. Genauer, E. 16-artist show is on today at Museum of Modern Art. *New York Herald Tribune*, 20 December 1959.
72. Jones, C.A. 1998. *Machine in the Studio: Constructing the Postwar American Artist.* University of Chicago Press. Chicago: pp. 407–408.
73. Lu, H.D. & A.W. Roe. 2008. Functional organization of color domains in V1 and V2 of macaque monkey revealed by optical imaging. *Cereb. Cortex* **18:** 516–533.
74. Gowing, L. 1979. *Matisse.* Oxford University Press. New York: p. 59.
75. Albers, J. 1975. *Interaction of Color Unabridged Text & Selected Plates.* Yale University Press. New Haven.